CONTENTS

UNRAVELING THE GPS MATRIX: TECHNOLOGY, SECURITY, AND GLOBAL IMPACT

By

Tony Yustein © 2024
https://thecode.wiki

INTRODUCTION

The Transformative Power of GPS in Modern Life

Imagine a world where finding your way home required nothing more than the stars above or a compass in hand. Today, thanks to the transformative power of Global Positioning System (GPS) technology, we navigate with pinpoint precision using devices as small as a smartwatch. GPS has become so integral to our daily lives that it's easy to overlook its profound influence on modern society. From guiding aircraft safely to their destinations to enabling location-based services on our smartphones, GPS has reshaped the way we live, work, and connect.

However, as we become more reliant on this invisible grid of satellites, understanding its operation and vulnerabilities becomes ever more critical. This book aims to illuminate the science behind GPS, its history, its applications, and the challenges of keeping it secure in a rapidly advancing technological landscape.

• •

A BRIEF HISTORY OF NAVIGATION SYSTEMS LEADING TO GPS

The journey to GPS did not begin with satellites but with ancient mariners charting their courses by the stars. Over centuries, human ingenuity gave rise to increasingly sophisticated navigation systems:

1. **Celestial Navigation**
   Early seafarers used the position of celestial bodies to guide their journeys. Tools like the sextant and nautical almanacs revolutionized oceanic navigation but remained reliant on clear skies and careful observation.

2. **The Magnetic Compass and Chronometer**
   The invention of the compass and marine chronometer marked a leap in navigation, providing more reliable methods for determining direction and position. However, these tools still required significant skill and manual calculations.

3. **The Radio and LORAN Era**
   With the advent of radio technology in the 20th century, navigation systems like LORAN (Long Range Navigation) allowed precise positioning using radio signals. These systems laid the groundwork for satellite-

based navigation.

4. **The Birth of GPS**

   In the late 20th century, the U.S. Department of Defense developed GPS as a military tool to ensure precise navigation for troops and equipment. The launch of the first NAVSTAR satellites in the 1970s marked the beginning of this revolutionary technology. By the 1990s, GPS became accessible for civilian use, transforming it into a global utility.

• •

THE IMPORTANCE OF UNDERSTANDING GPS SCIENCE, APPLICATIONS, AND SECURITY

The ubiquity of GPS in modern life is undeniable. Its applications extend far beyond navigation:

- **Logistics and Transportation**: GPS enables real-time tracking of goods, optimizing supply chains and improving delivery accuracy.

- **Emergency Services**: First responders use GPS to locate those in need quickly, saving countless lives.

- **Agriculture**: Precision farming relies on GPS for efficient resource management, enhancing productivity and sustainability.

- **Timing and Synchronization**: GPS provides the backbone for financial transactions, telecommunications, and power grid operations by offering highly accurate time synchronization.

Despite its benefits, GPS is not invulnerable. The system faces threats from jamming, spoofing, and cyberattacks, which can disrupt critical infrastructure and compromise safety. Furthermore, the geopolitical control of GPS and other Global

Navigation Satellite Systems (GNSS) raises questions about accessibility and security in a world increasingly dependent on location-based services.

Understanding how GPS works, its limitations, and how to protect it is essential for everyone—from policymakers and technologists to everyday users. This knowledge equips us to appreciate its benefits while safeguarding it against misuse and failure.

• •

BOOK STRUCTURE AND KEY TOPICS TO BE COVERED

This book, *"Navigating the Grid: The Science, Operation, and Security of GPS Systems,"* is divided into four parts to provide a comprehensive exploration of GPS technology:

1. **Part I: The Science and Operation of GPS Systems**
   Readers will learn the fundamental principles of GPS, including how satellites communicate with receivers, the physics behind signal transmission, and how errors are mitigated to achieve high precision. We'll also explore the global systems that complement GPS, such as GLONASS, Galileo, and BeiDou.

2. **Part II: Challenges and Threats to GPS Systems**
   This section delves into the vulnerabilities of GPS, such as signal jamming and spoofing. It examines the technical and practical challenges of maintaining accuracy and reliability, especially in environments like urban canyons or under heavy foliage.

3. **Part III: Global Implications and Emerging Trends**
   Here, we'll explore how GPS shapes geopolitics and global industries. Topics include international operators of GNSS, ethical considerations, and emerging technologies like quantum navigation that could redefine positioning systems.

4. **Part IV: Practical Implementation Guides**

The final section provides hands-on guidance for securing GPS systems against threats. From setting up anti-jamming equipment to testing anti-spoofing defenses, readers will gain actionable insights to protect critical infrastructure and enhance GPS resilience.

• •

CONCLUSION

GPS has revolutionized the modern world, but with great power comes great responsibility. By understanding the science, applications, and security of GPS systems, we can harness its potential while addressing its vulnerabilities. Whether you are a technologist, policymaker, or curious reader, this book offers a comprehensive guide to one of the most transformative technologies of our time.

Let us embark on this journey through the intricate world of GPS —its science, its impact, and its future.

CHAPTER 1: THE FUNDAMENTALS OF GPS TECHNOLOGY

The Global Positioning System (GPS) is a marvel of modern engineering, seamlessly connecting space-based satellites, ground control systems, and user devices to provide precise location, navigation, and timing information. To understand how GPS works, we must delve into its fundamental principles, signal components, and structural segments.

• •

PRINCIPLES OF TRILATERATION: HOW POSITION IS CALCULATED USING SATELLITES

At the heart of GPS lies the principle of **trilateration**, a geometric method for determining an unknown position based on its distance from three or more known reference points.

1. **Distance Measurement**

   Each GPS satellite continuously broadcasts a signal containing its precise location and the time the signal was sent. GPS receivers calculate their distance to each satellite by measuring the time it takes for the signal to arrive, multiplying this time by the speed of light.

   Distance = Time Delay × Speed of Light

2. **Trilateration in Two Dimensions**

   In a simplified 2D example, if the receiver is a known distance from three reference points (e.g., A, B, and C), the intersection of the three circles defined by these distances reveals the receiver's position.

3. **Expanding to Three Dimensions**

   In reality, GPS operates in three dimensions, requiring signals from at least four satellites to determine latitude, longitude, altitude, and time. The fourth satellite corrects for any clock discrepancies in the receiver.

4. **Error Sources and Corrections**
   Various factors—such as atmospheric delays, satellite clock errors, and multipath interference—can affect signal accuracy. GPS employs error-correction mechanisms, including:

   – **Differential GPS (DGPS):** Ground-based corrections for enhanced precision.

   – **Satellite-based augmentation systems (SBAS):** Such as WAAS or EGNOS.

• •

COMPONENTS OF GPS SIGNALS

Each GPS satellite transmits unique signals composed of three critical elements:

1. **Frequency Bands**
   GPS signals operate on specific radio frequencies to minimize interference:

   – **L1 Band (1575.42 MHz):** Primarily for civilian use, carrying the Coarse/Acquisition (C/A) code.

   – **L2 Band (1227.60 MHz):** Used by military systems, carrying the encrypted P(Y) code.

   – **L5 Band (1176.45 MHz):** For advanced civilian and safety-of-life applications.

2. **Pseudorandom Codes**
   Each satellite generates a unique pseudorandom noise (PRN) code, ensuring that signals from multiple satellites can be distinguished by the receiver. These codes include:

   – **C/A Code:** Openly accessible for civilian applications.

   – **P(Y) Code:** Encrypted for military use, providing higher accuracy and security.

3. **Navigation Messages**
   GPS signals also include navigation data, providing essential information about:

   – Satellite position and health (ephemeris data).

- Timing and orbital corrections.
- Atmospheric models for signal delay estimation.

. . . .

THE THREE SEGMENTS OF GPS

The GPS system consists of three interdependent segments: the **space segment**, the **control segment**, and the **user segment**. Each plays a vital role in delivering reliable and accurate positioning data.

Space Segment: Satellite Constellations

- The space segment comprises a constellation of at least **24 operational satellites** (plus spares) orbiting at an altitude of approximately **20,200 kilometers** in six orbital planes. These satellites ensure global coverage, with any point on Earth having access to signals from at least four satellites at all times.

- **Functions of GPS Satellites**:

    - Broadcasting timing signals and orbital data.

    - Monitoring their own health and functionality.

    - Resisting environmental factors like radiation and temperature extremes.

- **Advancements**:

    - The launch of **GPS III satellites** offers improved accuracy, anti-jamming capabilities, and a longer lifespan.

Control Segment: Ground Stations and Monitoring

- The control segment operates and maintains the GPS constellation through a network of **ground stations** distributed across the globe. It comprises:

- **Master Control Station (MCS):** Located at Schriever Space Force Base, Colorado, responsible for overall satellite control.

- **Monitor Stations:** Spread worldwide to track satellite signals and report anomalies.

- **Ground Antennas:** Used to communicate with satellites, sending updated data and instructions.

- **Core Functions:**

  - Continuously updating satellite ephemeris and clock data.

  - Ensuring synchronization across all satellites.

  - Detecting and correcting satellite errors to maintain system integrity.

User Segment: GPS Receivers

- The user segment consists of GPS receivers, which range from simple smartphone applications to highly specialized military or scientific devices. Receivers process satellite signals to determine:

  - **Position**: Latitude, longitude, and altitude.

  - **Velocity**: Speed and direction of movement.

  - **Time**: Highly accurate time synchronization.

- **Types of Receivers:**

  - **Consumer-grade receivers:** Found in smartphones, vehicles, and wearable devices.

  - **Survey-grade receivers:** Used in geodesy and engineering for centimeter-level accuracy.

  - **Military receivers:** Encrypted and designed to function in hostile environments.

- **Emerging Trends**:

  - Integration with multi-constellation GNSS systems for enhanced accuracy.

  - Incorporation of AI and machine learning to improve error correction.

• •

CONCLUSION

GPS technology is a seamless blend of physics, engineering, and data processing, with its foundation rooted in trilateration and precise signal transmission. The collaboration of the space, control, and user segments ensures the accuracy and reliability of this indispensable system. By understanding these fundamentals, we can better appreciate the sophistication and challenges of GPS technology.

CHAPTER 2: THE PHYSICS BEHIND GPS

The Global Positioning System (GPS) operates at the intersection of advanced physics and engineering. Its precision relies on the intricate interplay of satellite motion, timekeeping, and the behavior of electromagnetic signals. In this chapter, we will delve into the key physical principles that enable GPS, including satellite orbits, atomic clocks, relativity, and signal propagation.

• •

Satellite Orbits: Keplerian Motion and Geostationary vs. Geosynchronous Satellites

Keplerian Motion

- GPS satellites move in predictable orbits, governed by Kepler's laws of planetary motion:

    1. **First Law (Elliptical Orbits):** Satellites orbit the Earth in elliptical paths, with the Earth at one of the foci. However, GPS satellites are positioned in nearly circular orbits for stability and consistency.

    2. **Second Law (Equal Areas):** A satellite moves faster when closer to Earth (perigee) and slower when farther away (apogee), though for GPS satellites, this variance is minimal due to their near-circular orbits.

    3. **Third Law (Orbital Period):** The square of a satellite's orbital period is proportional to the cube of the semi-major axis of its orbit. GPS satellites

complete an orbit in approximately **12 hours**.

Geostationary vs. Geosynchronous Satellites

- **Geostationary Satellites:**

  – Positioned at an altitude of approximately **35,786 km**.

  – Remain fixed relative to a point on Earth's equator, ideal for applications like television broadcasting.

  – Not used in GPS, as they do not provide global coverage.

- **Geosynchronous Satellites:**

  – GPS satellites are in **geosynchronous orbits**, meaning their orbital period matches Earth's rotation but they are not fixed over one point.

  – Altitude: Approximately **20,200 km**.

  – This orbit ensures a consistent constellation that provides coverage to any point on Earth.

Orbital Planes and Coverage

- The GPS constellation consists of **24 satellites** distributed across **6 orbital planes**, each inclined at **55 degrees** to the equator.

- This configuration ensures global coverage, with a minimum of four satellites visible from any location on Earth.

• •

THE ROLE OF ATOMIC CLOCKS AND PRECISE TIMING IN GPS

Atomic Clocks in Satellites

- GPS satellites are equipped with highly precise atomic clocks that use the vibrations of cesium or rubidium atoms to measure time.

- **Accuracy:** These clocks drift by no more than **one nanosecond per day** (1 ns = 10^{-9} seconds).

Importance of Timing

- Precise timing is the cornerstone of GPS. A signal traveling at the speed of light (approximately 3×10^{8} m/s) takes just milliseconds to reach a receiver. A timing error of 1 microsecond could result in a positional error of **300 meters**.

- Each satellite broadcasts the exact time the signal is sent, enabling the receiver to calculate the distance to the satellite by measuring the time delay.

Synchronization

- To ensure synchronization:

  - Ground control stations regularly update satellite clocks.

  - GPS receivers correct for minor clock errors using data

from multiple satellites.

Time Standards

- GPS time is based on **Coordinated Universal Time (UTC)**, but it does not incorporate leap seconds, making it slightly ahead of UTC by a known offset.

· ·

RELATIVITY'S IMPACT ON GPS (TIME DILATION AND GRAVITATIONAL EFFECTS)

Einstein's theory of relativity plays a significant role in GPS accuracy, as both **special relativity** and **general relativity** affect satellite clocks.

Special Relativity: Time Dilation

- Satellites travel at high speeds (~14,000 km/h), causing their clocks to tick slower relative to stationary clocks on Earth due to time dilation.

- **Effect:** Clocks onboard satellites run approximately **7 microseconds slower per day**.

General Relativity: Gravitational Time Dilation

- GPS satellites are farther from Earth's gravitational field compared to ground-based clocks. According to general relativity, weaker gravity causes clocks to tick faster.

- **Effect:** Satellite clocks run approximately **45 microseconds faster per day**.

Net Effect

- Combining these two effects, satellite clocks run **38**

microseconds faster per day relative to Earth-based clocks.

- If uncorrected, this discrepancy would result in positional errors of several kilometers daily.

Corrections

- GPS satellites pre-adjust their clocks to account for relativistic effects before launch.

- Receivers incorporate relativistic corrections using models included in satellite signals.

• •

SIGNAL PROPAGATION AND THE SPEED OF LIGHT

Speed of Signal Transmission

- GPS signals travel at the speed of light in a vacuum (~3×10^8 m/s). However, the speed is slightly reduced as the signals pass through the Earth's atmosphere.

Atmospheric Effects

1. **Ionosphere:**

   - The ionosphere is a layer of charged particles that can delay GPS signals, causing variability in their speed.

   - Correction: Dual-frequency receivers can measure delays at two frequencies to calculate and mitigate ionospheric errors.

2. **Troposphere:**

   - The lower atmosphere introduces delays due to water vapor and air density.

   - Correction: Models of atmospheric conditions are used to estimate and reduce these delays.

Multipath Interference

- Signals can bounce off buildings, terrain, or other surfaces before reaching the receiver, creating errors.

- Advanced receivers use algorithms to identify and filter out multipath signals.

Error Budgets

- Atmospheric effects, clock errors, and multipath interference collectively create an **error budget** for GPS accuracy.

- Differential GPS and satellite-based augmentation systems (SBAS) help reduce these errors, improving accuracy from several meters to a few centimeters.

. . . .

CONCLUSION

The physics of GPS is a testament to humanity's ability to harness complex principles for practical applications. From the precision of atomic clocks to the relativistic corrections required for accurate timing, GPS embodies the seamless integration of theoretical physics and engineering. As we navigate the invisible grid of satellites orbiting high above, the science behind GPS ensures its reliability and accuracy.

CHAPTER 3: GLOBAL NAVIGATION SATELLITE SYSTEMS (GNSS)

While GPS is the most well-known navigation satellite system, it is part of a larger ecosystem of Global Navigation Satellite Systems (GNSS). These systems, operated by various nations, provide worldwide and regional navigation services for civilian, military, and commercial use. In this chapter, we will explore the major GNSS systems, their regional augmentation counterparts, and a comparative analysis of their accuracy, coverage, and reliability.

. . . .

OVERVIEW OF GLOBAL SYSTEMS

1. GPS (United States)

- **Full Name:** Global Positioning System

- **Operator:** United States Space Force

- **Established:** Initially developed in the 1970s for military use; declared fully operational in 1995.

- **Constellation:**

  – Consists of **31 satellites** (24 operational and spares) in medium Earth orbit (MEO) at approximately **20,200 km** altitude.

  – Distributed across six orbital planes inclined at **55 degrees**.

- **Capabilities:**

  – Provides global coverage for positioning, navigation, and timing (PNT).

  – Civilian signals (L1, L5) and encrypted military signals (M-code) offer high accuracy and resilience against interference.

- **Applications:**

  – Widely used in civilian sectors like transportation, agriculture, and emergency services.

  – Integral to U.S. military operations worldwide.

• •

2. GLONASS (Russia)

- **Full Name:** Global Navigation Satellite System

- **Operator:** Russian Aerospace Forces

- **Established:** Developed during the Soviet era in the 1980s; declared fully operational in 1996 (modernized in the 2000s).

- **Constellation:**

  – Comprises **24 operational satellites** in three orbital planes at **19,100 km** altitude.

  – Orbital inclination of **64.8 degrees** provides better coverage in high latitudes, particularly for Russia.

- **Capabilities:**

  – Provides PNT services with a focus on Russia and surrounding regions.

  – Offers an alternative to GPS for global users, though accuracy is slightly lower.

- **Applications:**

  – Used extensively within Russia for military and civilian purposes.

• •

3. Galileo (European Union)

- **Operator:** European Union Agency for the Space Programme (EUSPA)

- **Established:** First satellite launched in 2011; declared operational in 2016.

- **Constellation:**

  – Designed for **30 satellites** (24 operational and spares) in MEO at **23,222 km** altitude.

- – Inclination of **56 degrees** ensures global coverage.
- **Capabilities:**
  - – High-accuracy services with dual-frequency signals available for civilian users.
  - – Provides unique features like an integrity monitoring service for safety-critical applications.
- **Applications:**
  - – Promotes European independence from other GNSS systems.
  - – Widely used in the aviation, maritime, and IoT sectors.

. . . .

4. BeiDou (China)

- **Full Name:** BeiDou Navigation Satellite System (BDS)
- **Operator:** Chinese government
- **Established:** First operational in 2000 with regional coverage; global services available since 2020 (BDS-3).
- **Constellation:**
  - – **35 satellites**: 24 in MEO, 6 in geostationary orbit (GEO), and 5 in inclined geosynchronous orbit (IGSO).
  - – GEO and IGSO satellites enhance regional coverage over Asia-Pacific.
- **Capabilities:**
  - – Provides global PNT services with centimeter-level accuracy in China using augmentation.
  - – Dual-frequency signals reduce ionospheric errors.
- **Applications:**
  - – Used extensively in China's domestic infrastructure and military.

– Expanding its reach in international markets, particularly in the Belt and Road Initiative.

. . . .

5. NavIC (India)

- **Full Name:** Navigation with Indian Constellation
- **Operator:** Indian Space Research Organisation (ISRO)
- **Established:** Regional system operational since 2018.
- **Constellation:**

    – **7 satellites**: 3 in GEO and 4 in IGSO at approximately **36,000 km** altitude.

    – Focuses on the Indian subcontinent and surrounding regions.

- **Capabilities:**

    – Provides regional PNT services with high accuracy.

    – Dual-frequency signals reduce reliance on atmospheric corrections.

- **Applications:**

    – Supports India's strategic and civilian needs, such as disaster management and transportation.

. . . .

6. QZSS (Japan)

- **Full Name:** Quasi-Zenith Satellite System
- **Operator:** Japanese government
- **Established:** First operational satellite launched in 2010; fully operational in 2018.
- **Constellation:**

    – **4 satellites** in IGSO focused on the Asia-Pacific region, with plans to expand to 7 satellites.

- Unique orbits ensure that one satellite is always visible at a high elevation angle in Japan.

- **Capabilities:**

    - Enhances GPS accuracy in Japan through regional augmentation.

    - Provides additional signals compatible with GPS receivers.

- **Applications:**

    - Used extensively in Japanese industries, including agriculture, transportation, and disaster relief.

• •

REGIONAL AUGMENTATION SYSTEMS AND THEIR ROLES

Regional augmentation systems improve GNSS accuracy and reliability for safety-critical applications. Examples include:

1. **WAAS (Wide Area Augmentation System)**

   – Operated by the United States for aviation safety.

   – Provides corrections for GPS signals over North America.

2. **EGNOS (European Geostationary Navigation Overlay Service)**

   – Operated by the European Union.

   – Enhances Galileo and GPS signals for aviation and agriculture in Europe.

3. **MSAS (Multi-functional Satellite Augmentation System)**

   – Operated by Japan.

   – Improves GPS signals for aviation in the Asia-Pacific region.

4. **GAGAN (GPS Aided GEO Augmented Navigation)**

- Operated by India.
- Provides corrections for GPS and NavIC signals in the Indian subcontinent.

• •

COMPARISON OF GNSS SYSTEMS

1. Accuracy

- GPS: **~3–5 meters** for civilian users; **<1 meter** with augmentation.

- Galileo: **1 meter** for public services; **20 cm** for high-accuracy services.

- GLONASS: **4.5–7 meters** for civilian users.

- BeiDou: **~5 meters** globally; **centimeter-level** regionally with augmentation.

- NavIC: **10–20 meters**; **better accuracy** in the Indian region.

- QZSS: **Sub-meter accuracy** in Japan with GPS compatibility.

2. Coverage

- GPS, Galileo, GLONASS, and BeiDou: **Global coverage.**

- NavIC and QZSS: Regional focus with high accuracy in their respective areas.

3. Reliability

- GPS and Galileo: Highly reliable, with well-maintained constellations and redundancy.

- GLONASS: Reliable but vulnerable to geopolitical constraints.

- BeiDou: Increasing reliability with new satellites and international partnerships.

- NavIC and QZSS: Regional systems with strong performance within their coverage areas.

• •

CONCLUSION

GNSS systems collectively provide robust global and regional navigation solutions, with each system tailored to meet specific strategic, military, and civilian needs. Understanding these systems' capabilities, accuracy, and roles helps users select the best option for their applications while fostering collaboration between nations.

CHAPTER 4: HOW GPS DEVICES WORK

The functionality of GPS devices is a sophisticated combination of hardware, software, and algorithms designed to interpret satellite signals into precise positioning, velocity, and time data. This chapter delves into the anatomy of a GPS receiver, explores the algorithms that drive its calculations, and examines the various error sources and correction techniques that enhance its accuracy.

• •

ANATOMY OF A GPS RECEIVER: HARDWARE AND SOFTWARE

A GPS receiver is a highly specialized device engineered to capture, process, and interpret signals from satellites. It comprises several essential components:

Hardware Components

1. **Antenna**

   – The antenna captures the radio signals transmitted by GPS satellites.

   – Typically designed as a patch or helical antenna, it is optimized for L-band frequencies (e.g., L1: 1575.42 MHz, L2: 1227.60 MHz).

   – Antennas with multi-frequency capabilities (e.g., L5: 1176.45 MHz) provide enhanced accuracy by compensating for atmospheric distortions.

2. **RF Front-End**

   – Filters and amplifies the received satellite signals to make them suitable for further processing.

   – Converts high-frequency signals to intermediate frequencies for digital processing.

3. **Digital Signal Processor (DSP)**

   – Extracts pseudorandom noise (PRN) codes from the

signal to identify the transmitting satellite.

– Measures signal timing to compute the distance between the receiver and each satellite.

4. **Clock System**

– GPS receivers typically use quartz oscillators for timing, but these are less precise than the atomic clocks onboard satellites.

– Timing errors are corrected by comparing data from multiple satellites.

5. **Processor and Memory**

– A microprocessor runs the navigation algorithms and user interface software.

– Memory stores satellite almanac data, ephemeris data, and user settings.

Software Components

1. **Signal Acquisition and Tracking**

– The software locks onto visible satellites and continuously tracks their signals.

– Signal tracking ensures real-time data flow for dynamic applications.

2. **Navigation Algorithms**

– Algorithms calculate the receiver's position, velocity, and time (PVT).

– Implements error-correction methods to refine accuracy.

3. **User Interface**

– Provides information such as coordinates, maps, and directions.

– In advanced devices, integrates features like route

optimization and traffic updates.

• •

ALGORITHMS FOR POSITION, VELOCITY, AND TIME (PVT) CALCULATION

The core functionality of GPS lies in its ability to determine Position, Velocity, and Time (PVT) using satellite signals.

Position Calculation

- **Trilateration**:

  - The receiver determines its distance from at least four satellites.

  - These distances, combined with satellite location data, enable the calculation of the receiver's position in 3D space (latitude, longitude, and altitude).

Velocity Calculation

- Velocity is computed by measuring the Doppler shift of satellite signals, which indicates the relative speed of the satellite and receiver.

- Advanced receivers use carrier phase measurements for sub-meter velocity accuracy.

Time Synchronization

- The receiver's clock is synchronized with GPS system time using signals from multiple satellites.

- Timing synchronization is crucial for applications like

financial transactions and power grid management.

Error Correction Algorithms

- Algorithms identify and mitigate errors from atmospheric interference, signal reflection, and clock discrepancies.

- Integrated correction methods enhance the reliability of computed data.

• •

ERROR SOURCES AND CORRECTIONS

GPS signals are subject to various error sources that can degrade accuracy. To mitigate these, correction methods like Differential GPS (DGPS), Wide Area Augmentation System (WAAS), and Real-Time Kinematic (RTK) positioning are employed.

Error Sources

1. **Atmospheric Errors**

   – **Ionosphere:** Causes signal delays due to charged particles.

   – **Troposphere:** Affects signals due to temperature, pressure, and humidity.

2. **Satellite Clock Errors**

   – Deviations in satellite atomic clocks introduce timing inaccuracies.

3. **Multipath Interference**

   – Signals reflecting off surfaces before reaching the receiver can cause incorrect distance calculations.

4. **Orbital Errors**

   – Slight inaccuracies in satellite ephemeris data lead to positional deviations.

5. **Receiver Noise**

   – Internal noise in the receiver can distort signal processing.

• •

Correction Techniques

1. Differential GPS (DGPS)

- **Overview:**

  – DGPS uses a network of ground-based reference stations with precisely known locations to correct GPS signal errors.

- **How It Works:**

  – Reference stations measure satellite signal errors and transmit correction data to nearby GPS receivers.

- **Accuracy Improvement:**

  – DGPS enhances accuracy to within **1–3 meters**.

- **Applications:**

  – Used in maritime navigation, agriculture, and surveying.

• •

2. Wide Area Augmentation System (WAAS)

- **Overview:**

  – WAAS is a satellite-based augmentation system (SBAS) developed by the FAA to improve GPS accuracy and reliability for aviation.

- **How It Works:**

  – Ground reference stations across North America monitor GPS signals and send correction data to geostationary satellites.

  – These satellites relay corrections to WAAS-enabled receivers.

- **Accuracy Improvement:**

  – Reduces errors to less than **1 meter**.

- **Applications:**

  – Critical for aviation, ensuring safe landing approaches and en-route navigation.

• •

3. Real-Time Kinematic (RTK) Positioning

- **Overview:**

  – RTK uses carrier-phase measurements and real-time corrections from a base station to achieve centimeter-level accuracy.

- **How It Works:**

  – A base station with a fixed, known position continuously compares received GPS signals with expected values.

  – Corrections are sent to RTK-enabled receivers via radio, cellular, or internet links.

- **Accuracy Improvement:**

  – RTK achieves accuracy within **1–2 centimeters**.

- **Applications:**

  – Widely used in surveying, precision agriculture, and autonomous vehicles.

• •

CONCLUSION

GPS devices are engineering marvels, combining advanced hardware, sophisticated software, and mathematical algorithms to provide accurate positioning, velocity, and time data. Despite inherent error sources, techniques like DGPS, WAAS, and RTK have pushed GPS accuracy to unprecedented levels, enabling a wide range of applications across industries. Understanding the internal workings of GPS receivers highlights the incredible science and innovation behind this indispensable technology.

CHAPTER 5: APPLICATIONS OF GPS

GPS technology has transformed numerous aspects of modern life, enabling precise navigation, timing, and location tracking across civilian, military, and emerging technological domains. This chapter explores the vast array of applications, highlighting how GPS integrates into existing systems and drives innovation in new fields.

• •

CIVILIAN USES OF GPS

1. Navigation

- **Personal Navigation Devices (PNDs):**

  - GPS is central to devices such as car navigation systems and smartphone apps (e.g., Google Maps, Waze).

  - Features include turn-by-turn directions, traffic updates, and estimated time of arrival (ETA) calculations.

- **Maritime and Aviation Navigation:**

  - Ships rely on GPS for open-sea navigation, port approach, and collision avoidance.

  - Pilots use GPS for en-route navigation, approach guidance, and landing in low visibility conditions.

- **Recreational Activities:**

  - Outdoor enthusiasts use GPS devices for hiking, geocaching, and off-road navigation.

  - GPS is vital in water sports for route planning and safety.

• •

2. Mapping and Geographic Information Systems (GIS)

- **Surveying and Cartography:**

  - GPS streamlines the collection of geospatial data, enabling accurate map creation.

  - Real-Time Kinematic (RTK) positioning enhances survey accuracy to centimeter levels.

- **Urban Planning and Infrastructure Development:**

  - GPS data supports city planners in designing efficient transport networks, utilities, and public spaces.

  - Applications in disaster management include mapping vulnerable areas and guiding relief operations.

- **Environmental Monitoring:**

  - Scientists use GPS to track changes in ecosystems, glacier movements, and coastal erosion.

  - Wildlife tracking with GPS collars aids in conservation efforts and understanding animal behaviors.

. . . .

3. Agriculture

- **Precision Farming:**

  - Farmers use GPS-guided tractors for accurate planting, fertilizing, and harvesting.

  - Variable rate technology (VRT) leverages GPS data to optimize resource use, reducing costs and environmental impact.

- **Livestock Management:**

  - GPS collars track the movement of livestock, helping farmers monitor grazing patterns and prevent theft.

- **Weather and Soil Monitoring:**

  - GPS-integrated sensors provide real-time data on soil conditions and weather, enabling informed decision-making.

. . . .

4. Internet of Things (IoT)

- **Asset Tracking:**

  - GPS is a cornerstone of IoT solutions for tracking goods

in transit, improving supply chain visibility.

- **Smart Cities:**

    – GPS powers smart city initiatives, including real-time traffic management, public transport tracking, and autonomous waste collection systems.

- **Wearable Devices:**

    – Fitness trackers and smartwatches use GPS to monitor user activities, such as running routes and cycling distances.

· ·

MILITARY APPLICATIONS OF GPS

1. Precision Weaponry

- **Guided Missiles and Bombs:**

    – GPS ensures pinpoint accuracy in targeting, minimizing collateral damage in combat zones.

    – Examples include Joint Direct Attack Munitions (JDAM) and cruise missiles.

- **Artillery Systems:**

    – GPS-guided artillery shells achieve high precision, extending range and reducing munitions waste.

• •

2. Reconnaissance and Intelligence Gathering

- **Surveillance Systems:**

    – GPS aids in locating and tracking enemy movements using drones and other reconnaissance platforms.

- **Troop and Asset Tracking:**

    – Soldiers use GPS to navigate hostile terrains and coordinate troop movements.

    – Command centers rely on real-time GPS data for tactical decision-making.

• •

3. Logistics and Supply Chain Management

- **Supply Chain Optimization:**

  – Military supply chains depend on GPS for efficient delivery of equipment, fuel, and provisions.

- **Search and Rescue Operations:**

  – GPS assists in locating and extracting personnel from combat zones or disaster-stricken areas.

• •

EMERGING USES
OF GPS

1. Autonomous Vehicles

- **Self-Driving Cars:**

    - GPS provides real-time positioning data essential for route planning and navigation.

    - Integration with sensors and AI allows vehicles to adapt to dynamic environments and traffic conditions.

- **Fleet Management:**

    - Companies like Tesla and Waymo utilize GPS to monitor and optimize autonomous vehicle fleets.

· ·

2. Drone Navigation

- **Commercial Drones:**

    - GPS enables precise flight paths for drones used in delivery, agriculture, and inspections.

    - Features include geofencing to restrict drones to designated areas.

- **Search and Rescue:**

    - Drones equipped with GPS and cameras assist in locating stranded individuals in hard-to-reach areas.

- **Military Drones:**

    - UAVs (Unmanned Aerial Vehicles) like Predator

and Reaper rely on GPS for target acquisition and navigation.

• •

3. Precision Robotics

- **Construction Robotics:**

  – GPS-guided robots enhance efficiency in large-scale construction projects by automating tasks like excavation and grading.

- **Medical Robotics:**

  – In healthcare, robots equipped with GPS assist in logistics, such as delivering supplies within hospitals.

- **Industrial Automation:**

  – Robots use GPS to navigate warehouses and manufacturing facilities, optimizing workflows.

• •

CONCLUSION

The applications of GPS span diverse fields, from guiding a hiker in the wilderness to coordinating military operations. As emerging technologies like autonomous vehicles and precision robotics continue to develop, GPS will remain a foundational element, driving innovation and connectivity. The versatility and reliability of GPS underscore its indispensable role in shaping modern life.

CHAPTER 6: THE ACCURACY AND LIMITATIONS OF GPS

GPS technology delivers extraordinary precision in determining position, velocity, and time (PVT). However, it is not immune to errors and limitations. Factors like atmospheric conditions, environmental obstacles, and inherent system inaccuracies can impact GPS performance. This chapter explores these challenges in detail and discusses the science behind these errors and their mitigation techniques.

• •

1. ATMOSPHERIC INTERFERENCE

The Earth's atmosphere significantly affects the propagation of GPS signals, introducing delays and inaccuracies.

Ionospheric Delays

- **Nature of the Ionosphere:**

    - The ionosphere is a layer of the atmosphere, located approximately **50 to 1,000 kilometers above the Earth's surface**, filled with charged particles (ions) created by solar radiation.

    - These ions interact with GPS signals, causing changes in signal speed and direction.

- **Impact on GPS Accuracy:**

    - Ionospheric interference can cause delays of up to **10–15 meters** in calculated positions.

    - Variability increases during periods of high solar activity or in equatorial regions where ionization is stronger.

- **Mitigation Techniques:**

    - **Dual-Frequency Receivers:** Measure delays on two different frequencies (e.g., L1 and L2) to calculate and correct ionospheric delays.

    - **Ionospheric Models:** Ground stations generate correction models, which are transmitted to receivers as part of the GPS navigation message.

• •

Tropospheric Delays

- **Nature of the Troposphere:**

  – The troposphere is the lowest layer of the atmosphere, extending up to 10–20 kilometers above the Earth's surface, where weather phenomena occur.

  – Unlike the ionosphere, the troposphere does not ionize signals but causes delays due to variations in temperature, pressure, and humidity.

- **Impact on GPS Accuracy:**

  – Tropospheric delays are generally smaller than ionospheric delays but can still introduce errors of **1–5 meters**.

- **Mitigation Techniques:**

  – **Tropospheric Models:** Predict delays based on meteorological data and incorporate corrections into GPS calculations.

  – **Onboard Sensors:** Advanced receivers equipped with barometers and thermometers adjust for local atmospheric conditions.

• •

2. MULTIPATH EFFECTS

Multipath effects occur when GPS signals reflect off surfaces before reaching the receiver, creating delays and inaccuracies.

Nature of Multipath Interference

- Reflective surfaces, such as buildings, water bodies, and terrain, can scatter GPS signals, leading to:

    - **Longer Signal Paths:** The reflected signal takes a longer path to the receiver, resulting in overestimated distances.

    - **Signal Degradation:** Reflected signals interfere with direct signals, reducing the clarity of the received signal.

Impact on GPS Accuracy

- Multipath effects are most significant in urban and indoor environments, where reflective surfaces are abundant.

- Errors caused by multipath interference can range from **a few meters to tens of meters**, depending on the environment.

Mitigation Techniques

- **Antenna Design:**

    - High-quality antennas with multipath rejection capabilities (e.g., choke-ring antennas) minimize interference from reflected signals.

- **Signal Processing:**

    - Advanced receivers use algorithms to distinguish

between direct and reflected signals by analyzing their arrival times and angles.

- **Environmental Adaptation:**

    – In areas prone to multipath effects, combining GPS with alternative navigation systems (e.g., inertial navigation systems) can enhance accuracy.

· ·

3. CLOCK AND EPHEMERIS ERRORS

Clock Errors

- **Nature of Satellite Clocks:**

    – GPS satellites rely on atomic clocks, which are extremely accurate but not perfect. Minor deviations in these clocks can introduce timing errors.

- **Impact on GPS Accuracy:**

    – A timing error of **1 nanosecond** can result in a positional error of **30 centimeters**.

- **Mitigation Techniques:**

    – **Ground Control Corrections:** Ground stations monitor satellite clocks and transmit corrections to receivers.

    – **Receiver Algorithms:** Receivers compare data from multiple satellites to identify and correct clock discrepancies.

• •

Ephemeris Errors

- **Nature of Ephemeris Data:**

    – Ephemeris data describes the precise orbital positions of satellites, which are subject to slight deviations due to gravitational forces and other factors.

- **Impact on GPS Accuracy:**

    – Errors in ephemeris data can cause positional

inaccuracies of **2–5 meters**.

- **Mitigation Techniques:**

  – Ground control stations continuously track satellite positions and update ephemeris data in navigation messages.

  – Receivers use real-time corrections provided by augmentation systems like WAAS or EGNOS.

• •

4. URBAN CANYONS AND ENVIRONMENTAL OBSTRUCTIONS

Urban environments and natural obstacles pose significant challenges to GPS signal reception and accuracy.

Urban Canyons

- **Nature of Urban Canyons:**

    - Tall buildings in cities create narrow pathways for GPS signals, causing signal blockages and multipath interference.

- **Impact on GPS Accuracy:**

    - Reduced satellite visibility increases the receiver's reliance on fewer signals, leading to degraded performance.

    - Errors can exceed **50 meters** in dense urban environments.

- **Mitigation Techniques:**

    - **Assisted GPS (A-GPS):** Combines GPS data with information from cellular networks to improve accuracy in urban areas.

    - **Dead Reckoning:** Uses inertial sensors to estimate position when GPS signals are unreliable.

• •

Environmental Obstructions

- **Nature of Obstructions:**

  – Dense foliage, mountains, tunnels, and water bodies can block or scatter GPS signals, reducing visibility and accuracy.

- **Impact on GPS Accuracy:**

  – Loss of line-of-sight to satellites significantly impacts positioning, particularly in remote or forested areas.

- **Mitigation Techniques:**

  – **Multi-Constellation Receivers:** Use signals from multiple GNSS systems (e.g., Galileo, GLONASS, BeiDou) to maintain accuracy.

  – **Hybrid Navigation:** Integrates GPS with other technologies like inertial navigation systems or LIDAR for continuous operation.

• •

CONCLUSION

GPS is an extraordinary technology, but it is not without limitations. Atmospheric interference, multipath effects, clock and ephemeris errors, and environmental obstructions all contribute to inaccuracies. However, advancements in hardware, software, and augmentation systems have significantly mitigated these challenges, enabling GPS to remain a cornerstone of modern navigation. Understanding these limitations empowers users to adopt strategies and technologies to maximize GPS performance.

CHAPTER 7:
SECURITY THREATS
TO GPS SYSTEMS

As indispensable as GPS is for modern life, it is not immune to security vulnerabilities. From deliberate interference to cyberattacks, threats to GPS can disrupt critical infrastructure, military operations, and civilian activities. This chapter explores the various security threats to GPS systems, their technical mechanisms, and real-world examples of their impact.

• •

1. JAMMING: TYPES, TOOLS, AND REAL-WORLD EXAMPLES

Nature of GPS Jamming

- **Definition:** Jamming involves emitting radio signals at the same frequency as GPS to overpower or disrupt legitimate satellite signals.

- **Mechanism:** GPS receivers rely on weak satellite signals, which can be easily overwhelmed by stronger, localized interference.

. . . .

Types of Jamming

1. **Barrage Jamming**

   – Emits a continuous broad-spectrum signal across GPS frequencies.

   – Disrupts all GPS operations in the affected area.

2. **Spot Jamming**

   – Focuses interference on a specific frequency (e.g., L1 or L2).

   – More energy-efficient but still effective in disabling GPS functionality.

3. **Sweep Jamming**

   – Continuously shifts the jamming signal across

frequencies.

– Prevents receivers from locking onto satellite signals.

4. Pulse Jamming

– Sends high-power, intermittent pulses to degrade GPS signals.

– Disrupts synchronization, particularly for timing-dependent applications.

. . . .

Tools Used for Jamming

• **Handheld Jammers:**

– Portable devices available on the black market, often used for local interference.

• **Military-Grade Jammers:**

– High-power systems capable of disrupting GPS over large areas, often mounted on vehicles or drones.

• **Software-Defined Radios (SDRs):**

– Highly customizable tools that can simulate complex jamming signals.

. . . .

Real-World Examples

• **North Korea (2016):**

– Conducted widespread GPS jamming targeting South Korea, affecting aviation, shipping, and cellular networks.

• **Black Sea Incident (2017):**

– Ships in the Black Sea reported GPS disruptions, suspected to be Russian military jamming.

• **Commercial Jamming:**

–   Truck drivers in the U.S. have used jammers to disable GPS tracking, inadvertently interfering with air traffic control systems.

• •

2. SPOOFING: TECHNIQUES AND IMPLICATIONS

Nature of GPS Spoofing

- **Definition:** Spoofing involves transmitting fake GPS signals to mislead receivers into calculating false positions or times.

- **Mechanism:** Spoofed signals mimic legitimate satellite signals but are stronger, causing receivers to lock onto them instead of authentic signals.

• •

Techniques of Spoofing

1. **Simple Spoofing**

   – Broadcasts a fixed false signal to confuse receivers.

   – Effective in causing navigational errors for basic GPS devices.

2. **Sophisticated Spoofing**

   – Sends dynamic, realistic signals that replicate satellite movements.

   – Can manipulate receivers into believing they are at a specific location.

3. **Meaconing**

   – Captures legitimate GPS signals and re-transmits them with delays to mislead receivers.

• •

Implications of Spoofing

- **Civilian Risks:**

  - Misguiding ships, aircraft, and autonomous vehicles can lead to accidents and economic losses.

- **Military Threats:**

  - Spoofing can disrupt precision-guided munitions or create navigational chaos during operations.

- **Critical Infrastructure:**

  - Spoofing of timing signals can compromise power grids, financial systems, and telecommunications.

• •

Real-World Examples

- **Iranian Capture of U.S. Drone (2011):**

  - Iran reportedly spoofed GPS signals to redirect a U.S. RQ-170 Sentinel drone into Iranian airspace.

- **Black Sea Spoofing (2017):**

  - Ships reported fake GPS locations miles inland, suspected to be a Russian spoofing operation.

- **China (2019):**

  - Spoofing was used to misdirect ships near ports, believed to be part of anti-surveillance measures.

• •

3. CYBER THREATS TO THE CONTROL SEGMENT

Nature of Cyber Threats

- The GPS control segment, comprising ground stations and communication networks, is vulnerable to cyberattacks targeting its infrastructure.

. . . .

Types of Cyber Threats

1. **Hacking**

   – Attackers gain unauthorized access to ground control systems, potentially altering satellite commands or navigation data.

2. **Data Interception**

   – Hackers intercept communication between satellites and ground stations, compromising data integrity.

3. **Denial-of-Service (DoS) Attacks**

   – Overwhelms ground systems with excessive requests, disrupting updates to satellites.

. . . .

Implications of Cyber Threats

- **Satellite Manipulation:**

   – Altering satellite orbits or disabling systems could

disrupt global navigation.

- **Data Corruption:**

  – Injecting false data into navigation messages would degrade user trust in GPS.

- **Service Outages:**

  – Cyberattacks on control centers could lead to prolonged GPS service interruptions.

• •

Real-World Examples

- **U.S. Satellite Hacking Attempt (2014):**

  – A suspected Chinese cyberattack targeted NASA's ground stations, raising concerns about GNSS vulnerabilities.

- **Galileo Outage (2019):**

  – A technical error, compounded by cybersecurity concerns, caused a week-long disruption in Galileo services.

• •

4. CASE STUDIES: MAJOR DISRUPTIONS CAUSED BY GPS VULNERABILITIES

Case Study 1: North Korea's Jamming Campaign

- **Incident:** Between 2010 and 2016, North Korea repeatedly jammed GPS signals targeting South Korea.

- **Impact:**

  – Disrupted navigation for thousands of flights and ships.

  – Forced South Korea to strengthen its reliance on alternative systems, including eLORAN.

• •

Case Study 2: Drone Redirection by Iran

- **Incident:** Iran reportedly spoofed GPS signals in 2011 to capture a U.S. RQ-170 drone.

- **Impact:**

  – Highlighted the vulnerability of unmanned aerial systems to spoofing attacks.

  – Accelerated U.S. efforts to enhance GPS security for military assets.

• •

Case Study 3: Galileo Service Outage

- **Incident:** A technical failure in Galileo's ground control system led to a week-long outage in 2019.

- **Impact:**

  – Undermined confidence in Galileo as a GPS alternative.

  – Prompted revisions in cybersecurity protocols for GNSS systems.

• •

CONCLUSION

The reliance on GPS across civilian, military, and critical infrastructure sectors underscores the need to address its vulnerabilities. Jamming, spoofing, and cyberattacks pose significant risks, with real-world examples demonstrating their potential for disruption. Strengthening GPS security through advanced technologies, robust protocols, and international cooperation is essential to ensure its resilience in an increasingly connected world.

CHAPTER 8: SCRAMBLING GPS SIGNALS: PRACTICAL TECHNIQUES

Scrambling GPS signals involves deliberate manipulation or interference with GPS transmissions to distort, block, or obfuscate their utility. While often associated with security and military applications, understanding the technical, ethical, and legal implications of GPS scrambling is essential for its responsible use. This chapter provides a comprehensive overview of scrambling techniques, tools, controlled experimental setups, and real-world case studies.

• •

1. UNDERSTANDING SCRAMBLING

How and Why Signals Are Scrambled

Scrambling GPS signals disrupts their transmission or reception, rendering them unreliable or unusable. Scrambling serves several purposes:

- **Military Applications:**

    - Protecting operations from adversary surveillance or targeting.

    - Creating GPS-denied environments for secure maneuvers or combat scenarios.

- **Testing and Research:**

    - Evaluating GPS system vulnerabilities and resilience.

    - Studying alternative navigation systems under GPS-denied conditions.

- **Civilian Use Cases:**

    - Rarely justifiable but may include controlled experiments or protection of sensitive facilities from unauthorized tracking.

Technical Mechanism:
- Scrambling targets the frequency, code, or timing of GPS signals, exploiting the weak power of satellite transmissions. Techniques include:

    - Overwhelming the signal with noise (jamming).

–   Manipulating or mimicking signal characteristics (spoofing).

· ·

Ethical and Legal Considerations

While scrambling can have legitimate purposes, its misuse poses significant risks and is often regulated or prohibited:

1. **Ethical Implications:**

–   **Civilian Risks:** Disrupting GPS signals can jeopardize public safety, especially in aviation, maritime navigation, and emergency services.

–   **Privacy Concerns:** The technology could be used to mask illicit activities or evade lawful tracking.

2. **Legal Restrictions:**

–   **International Regulations:** Many countries, including the U.S. and EU member states, prohibit unauthorized interference with GPS or GNSS signals.

–   **Licensing Requirements:** Military and research scrambling often requires government authorization and adherence to strict protocols.

–   **Penalties for Misuse:** Violations can result in fines, imprisonment, or both.

· ·

2. METHODS AND TOOLS

Signal Generators and Low-Power Jammers

1. **Signal Generators:**

   – Devices capable of producing customized GPS-like signals for testing purposes.

   – Applications:

     • Simulating various satellite signal scenarios.

     • Testing receiver resilience in controlled environments.

   – **Example Tools:**

     • Spirent GNSS simulators.

     • Rohde & Schwarz signal generators.

2. **Low-Power Jammers:**

   – Emitters designed to interfere with GPS signals over limited ranges.

   – Applications:

     • Preventing GPS tracking in specific areas.

     • Creating GPS-denied environments for training exercises.

   – **Example Tools:**

     • Portable jammers (regulated for testing in

shielded environments).

• •

Software-Defined Radio (SDR) Tools for Signal Manipulation

1. **Overview of SDR:**

   – SDRs are flexible hardware devices that use software to define radio frequencies and signal characteristics.

   – Ideal for developing custom GPS signal manipulations and scrambling techniques.

2. **Applications:**

   – Generating spoofed GPS signals.

   – Simulating interference patterns for testing.

   – Experimenting with advanced jamming techniques.

3. **Popular SDR Tools:**

   – **HackRF One:** Versatile SDR platform for signal generation and analysis.

   – **BladeRF:** High-performance SDR for GNSS research.

   – **USRP (Universal Software Radio Peripheral):** A professional-grade tool for complex GNSS experimentation.

4. **Software Platforms:**

   – GNURadio: Open-source software for building SDR applications.

   – MATLAB: A robust platform for simulating and analyzing GPS scrambling techniques.

• •

3. CONTROLLED EXPERIMENTS

Setting Up a Legal Test Environment

1. **Designing a Secure Setup:**

   – Conduct tests in shielded environments, such as anechoic chambers or remote locations, to avoid unintentional interference.

   – Use low-power jammers or SDRs configured for localized, controlled impact.

2. **Obtaining Permissions:**

   – Apply for legal permits from relevant authorities (e.g., FCC in the U.S., Ofcom in the U.K.).

   – Follow strict documentation and reporting protocols to ensure compliance.

3. **Equipment Requirements:**

   – GPS receivers for testing impact.

   – Signal generators or SDRs for controlled interference.

   – Spectrum analyzers to monitor signal behavior.

• •

Step-by-Step Simulation of GPS Interference

1. **Preparation:**

   – Define the test objective (e.g., evaluating receiver resilience, studying signal disruption patterns).

– Configure the environment to isolate signals and minimize unintended effects.

2. **Signal Generation:**

– Use a signal generator or SDR to produce interference signals.

– Experiment with different jamming or spoofing techniques (barrage, sweep, or pulse jamming).

3. **Data Collection:**

– Monitor the impact on GPS receivers.

– Record data on signal strength, positional errors, and recovery time.

4. **Analysis:**

– Compare results against baseline GPS performance.

– Identify weaknesses or vulnerabilities in receiver design.

• •

Case Studies of Scrambling in Military and Research Contexts

1. **Military Use Cases:**

– **Operation Desert Storm (1991):**

• U.S. forces used GPS scrambling to prevent Iraqi forces from exploiting civilian GPS signals.

– **Modern Exercises:**

• NATO and allied forces conduct regular GPS-jamming drills to prepare for electronic warfare scenarios.

2. **Research Experiments:**

– **Stanford University GPS Laboratory:**

• Investigated the effects of jamming and spoofing

on autonomous vehicle navigation.

- **European GNSS Research:**
    - Projects like STRIKE3 study GNSS interference for resilience enhancement.

3. **Civilian Implications:**

- **Tokyo Olympics (2021):**
    - Extensive testing of GPS jamming to protect drone operations from unauthorized interference.

• •

CONCLUSION

Scrambling GPS signals is a powerful tool for enhancing system resilience and preparing for potential threats. However, it comes with significant ethical and legal responsibilities. By understanding the techniques, tools, and controlled experimentation methods, professionals can contribute to GPS research and security without compromising public safety or violating regulations. Responsible use of these technologies ensures that they remain a valuable resource for safeguarding GPS systems.

CHAPTER 9: PROTECTING GPS SIGNALS: PRACTICAL TECHNIQUES

As GPS becomes increasingly vital to civilian, military, and industrial operations, protecting its signals from interference, spoofing, and other vulnerabilities is a critical priority. This chapter examines practical techniques for defending GPS systems, exploring anti-jamming and anti-spoofing methods, physical and technological solutions, and implementation tutorials for real-world applications.

• •

1. ANTI-JAMMING METHODS

Jamming occurs when malicious actors or environmental factors overwhelm GPS signals, disrupting receiver performance. Anti-jamming techniques aim to detect, mitigate, and prevent such interference.

Adaptive Antennas and Phased Arrays

- **Adaptive Antennas:**

  - These antennas dynamically adjust their reception patterns to reject interfering signals.

  - Techniques include:

    - **Null Steering:** Reduces the sensitivity of the antenna in the direction of jamming sources.

    - **Beamforming:** Enhances signal reception from desired directions while suppressing noise.

- **Phased Arrays:**

  - Composed of multiple antenna elements that work together to form highly directional beams.

  - Benefits:

    - Increased resilience against jamming.

    - Capability to track multiple satellites simultaneously, even in contested environments.

  - Applications:

    - Used in advanced military systems and critical

infrastructure.

Spread-Spectrum Technology

- **Overview:**

  – Spread-spectrum techniques spread GPS signals over a wide range of frequencies, making them harder to isolate and jam.

- **Types:**

  – **Frequency Hopping:** The signal rapidly switches frequencies, preventing consistent jamming.

  – **Direct Sequence Spread Spectrum (DSSS):** Encodes the signal with a pseudorandom code, spreading it across a wider bandwidth.

- **Applications:**

  – Civilian GPS signals, like those on the L1 frequency, use DSSS.

  – Military signals (e.g., M-code) incorporate advanced spread-spectrum techniques.

• •

2. ANTI-SPOOFING METHODS

Spoofing involves broadcasting counterfeit GPS signals to deceive receivers. Anti-spoofing measures focus on detecting and countering these fake signals.

Cryptographic Authentication: Public and Private Key Infrastructure

- **Overview:**

    - Cryptographic authentication ensures the authenticity of GPS signals, preventing spoofed signals from being accepted.

- **Methods:**

    - **Public Key Infrastructure (PKI):**

        - Uses asymmetric cryptography to verify the source of GPS signals.

        - Public keys are shared openly, while private keys remain secure with the satellite.

    - **Digital Signatures:**

        - Signals are signed with cryptographic hashes that can be verified by receivers.

- **Applications:**

    - Military GPS signals, such as P(Y)-code and M-code, employ encryption and authentication.

    - Emerging civilian applications are exploring open

cryptographic solutions for enhanced security.

Signal Monitoring and Real-Time Validation

- **Overview:**

  - Continuous monitoring of GPS signals for anomalies is critical in detecting spoofing attempts.

- **Techniques:**

  - **Signal Strength Analysis:** Identifies unusually strong or weak signals as potential spoofing attempts.

  - **Time and Frequency Consistency Checks:** Verifies that signals align with expected satellite orbits and timing.

  - **Cross-Referencing with Other GNSS:** Confirms positions using multiple satellite systems (e.g., GPS, Galileo, GLONASS).

- **Applications:**

  - Integrated into advanced GPS receivers and GNSS monitoring stations.

• •

3. PHYSICAL AND TECHNOLOGICAL SOLUTIONS

Beyond software and algorithmic defenses, physical and technological measures enhance GPS resilience against threats.

Hardened Satellite Designs

- **Radiation-Hardened Components:**

    - Protect satellites from space weather, such as solar flares, that can affect signal integrity.

- **Anti-Jamming and Anti-Spoofing Transmitters:**

    - Satellites can transmit encrypted and high-power signals to resist interference.

- **Redundancy:**

    - Modern satellites are equipped with multiple signal generators and communication systems to ensure uninterrupted service.

Shielding Receivers in Critical Applications

- **Overview:**

    - Shielding prevents unauthorized access to GPS signals and mitigates the impact of jamming or spoofing.

- **Methods:**

    - **Faraday Cages:** Enclose receivers in conductive materials to block external electromagnetic interference.

- **Underground or Shielded Bunkers:** Protect critical receivers in military or government installations.

• **Applications:**

- Used in military bases, financial data centers, and critical infrastructure facilities.

• •

4. TUTORIALS: TESTING AND IMPLEMENTING ANTI-JAMMING AND SPOOFING DEFENSES

Testing Anti-Jamming Measures

1. **Setup:**

   – Deploy a GPS receiver and jammer in a controlled, shielded environment.

   – Introduce jamming signals at varying power levels and frequencies.

2. **Procedure:**

   – Measure the receiver's performance (e.g., signal strength, accuracy) before and during jamming.

   – Test adaptive antennas or phased arrays by simulating directional interference.

3. **Analysis:**

   – Evaluate the effectiveness of the anti-jamming technology and identify areas for improvement.

Testing Anti-Spoofing Measures

1. **Setup:**

- Use a software-defined radio (SDR) to generate spoofed signals.

- Deploy a GPS receiver equipped with anti-spoofing algorithms.

2. **Procedure:**

- Simulate various spoofing scenarios, including fixed and dynamic false signals.

- Monitor the receiver's ability to detect and reject spoofed signals.

3. **Analysis:**

- Assess the robustness of cryptographic authentication and real-time validation.

Implementing Defenses in Real-World Applications

1. **Receiver Configuration:**

- Enable multi-frequency and multi-GNSS support for redundancy.

- Activate built-in anti-jamming and anti-spoofing features.

2. **Integration with Monitoring Systems:**

- Connect receivers to GNSS monitoring networks for real-time alerts and updates.

3. **Regular Updates:**

- Maintain firmware and software updates to address emerging threats.

• •

CONCLUSION

Protecting GPS signals is vital for ensuring the reliability and safety of systems that depend on accurate navigation and timing. By employing advanced anti-jamming and anti-spoofing methods, along with physical and technological solutions, users can defend against interference and maintain GPS integrity. Testing and implementing these defenses in controlled and real-world environments enhances resilience, safeguarding critical infrastructure and operations from evolving threats.

CHAPTER 10: INTERNATIONAL OPERATORS AND GOVERNANCE

Global Navigation Satellite Systems (GNSS) are operated by multiple nations, each with its own infrastructure, policies, and geopolitical considerations. As GPS has evolved into an indispensable global utility, its governance has required regulatory oversight, international cooperation, and strategic management. This chapter explores the international operators of GNSS, regulatory frameworks, policies on signal use, and the geopolitical dynamics surrounding the militarization of satellite navigation.

. . . .

1. OVERVIEW OF COUNTRIES OPERATING GNSS SYSTEMS

United States: Global Positioning System (GPS)

- **Operator:** United States Space Force.

- **Coverage:** Global.

- **Features:**

  - Highly reliable with 31 operational satellites providing dual-frequency signals (L1, L2, and L5).

  - Open civilian signals and encrypted military signals.

- **Strategic Importance:**

  - Dominates the global GNSS landscape, supporting military, civilian, and commercial applications.

  - Integral to U.S. military operations, enabling precision weaponry and secure communications.

• •

Russia: Global Navigation Satellite System (GLONASS)

- **Operator:** Russian Aerospace Forces.

- **Coverage:** Global.

- **Features:**

- 24 operational satellites with robust performance in high-latitude regions.

- Alternative to GPS with slightly lower accuracy for civilian applications.

- **Strategic Importance:**

  - Offers Russia independence from GPS, particularly in defense and critical infrastructure.

  - Used extensively by Russian military and allied countries.

• •

European Union: Galileo

- **Operator:** European Union Agency for the Space Programme (EUSPA).

- **Coverage:** Global.

- **Features:**

  - 24 operational satellites with plans for additional launches.

  - High accuracy for civilian applications, with encrypted services for governmental use.

- **Strategic Importance:**

  - Provides the EU with GNSS autonomy, reducing reliance on GPS.

  - Enhances resilience for aviation, maritime, and IoT applications.

• •

China: BeiDou Navigation Satellite System (BDS)

- **Operator:** Chinese government.

- **Coverage:** Global (with enhanced regional services in Asia-Pacific).

- **Features:**

  - 35 satellites, including geostationary (GEO), inclined geosynchronous (IGSO), and medium Earth orbit (MEO) satellites.

  - Centimeter-level accuracy regionally with augmentation.

- **Strategic Importance:**

  - Supports China's military and civilian ambitions, including the Belt and Road Initiative.

  - Expanding influence by promoting BeiDou adoption in developing countries.

- -

India: Navigation with Indian Constellation (NavIC)

- **Operator:** Indian Space Research Organisation (ISRO).

- **Coverage:** Regional (Indian subcontinent and surrounding regions).

- **Features:**

  - 7 satellites in geosynchronous and inclined geosynchronous orbits.

  - Focus on regional autonomy and high-accuracy services.

- **Strategic Importance:**

  - Reduces India's dependence on foreign GNSS for defense and disaster management.

  - Expands applications in agriculture, logistics, and aviation.

- -

Japan: Quasi-Zenith Satellite System (QZSS)

- **Operator:** Japanese government.

- **Coverage:** Regional (Asia-Pacific, with a focus on Japan).
- **Features:**
  - 4 operational satellites in inclined geosynchronous orbits.
  - Provides GPS augmentation for enhanced accuracy.
- **Strategic Importance:**
  - Improves resilience for Japan's transportation, agriculture, and disaster response systems.
  - Facilitates regional leadership in satellite navigation technology.

• •

2. REGULATORY BODIES

International Telecommunication Union (ITU)

- **Role:**

    – Allocates frequency bands for GNSS signals to prevent interference between systems.

    – Ensures international coordination for the use of orbital slots and frequencies.

- **Significance:**

    – Balances the needs of different GNSS operators to maintain global signal integrity.

    – Monitors compliance with international agreements on spectrum use.

• •

Other Regulatory Bodies

- **European Space Agency (ESA):** Manages Galileo's development and operations.

- **United Nations Office for Outer Space Affairs (UNOOSA):** Oversees international cooperation on satellite technology, including GNSS.

- **National Telecommunications Authorities:**

    – Regulate signal usage within national territories (e.g., FCC in the U.S., Ofcom in the U.K.).

3. POLICIES ON SIGNAL USE, INTERFERENCE, AND EXPORT CONTROL

Signal Use

- **Open Signals:**

    - Available for civilian applications, typically on frequencies like L1 and L5.

    - Examples: GPS Standard Positioning Service (SPS), Galileo Open Service (OS).

- **Encrypted Signals:**

    - Restricted to authorized military or governmental users.

    - Examples: GPS M-code, Galileo Public Regulated Service (PRS).

• •

Interference Policies

- **International Treaties:**

    - Prohibit deliberate jamming or spoofing of GNSS signals in peacetime.

    - Regulated under agreements like the Outer Space Treaty (1967).

- **National Regulations:**

    – Many countries impose strict penalties for unauthorized interference with GNSS signals.

• •

Export Control

- **GNSS Technology Export:**

    – Advanced GNSS hardware and software are often subject to export restrictions to prevent misuse.

    – Example: The U.S. International Traffic in Arms Regulations (ITAR) govern the export of GPS technology.

- **Signal Access Control:**

    – Some GNSS operators restrict access to encrypted signals to allies or strategic partners.

• •

4. GEOPOLITICAL DYNAMICS AND THE MILITARIZATION OF GNSS

Geopolitical Competition

- **U.S. Dominance:**

    - GPS's global reach has cemented U.S. leadership in GNSS technology.

    - Other nations, like China and Russia, seek to reduce dependence on GPS for strategic autonomy.

- **Emerging Players:**

    - Regional systems like NavIC and QZSS reflect a growing trend toward localized GNSS solutions.

• •

Militarization of GNSS

- **Precision Warfare:**

    - GNSS enables precision-guided munitions, real-time troop tracking, and reconnaissance.

- **Electronic Warfare:**

    - Jamming and spoofing capabilities are increasingly used in military conflicts to disrupt enemy navigation.

- **Arms Race:**

  – Nations are investing in anti-GNSS technologies, such as signal denial systems and counter-jamming tools.

· ·

International Cooperation and Tensions

- **Cooperation:**

  – GNSS operators collaborate on interoperability, allowing receivers to use multiple systems for enhanced accuracy.

  – Example: GPS-Galileo compatibility agreements.

- **Tensions:**

  – GNSS systems are also tools of geopolitical influence, with countries using their systems to strengthen alliances or gain leverage in trade and security.

· ·

CONCLUSION

The operation and governance of GNSS systems reflect a balance of international cooperation and geopolitical competition. As more nations develop and deploy satellite navigation systems, the regulatory frameworks provided by bodies like the ITU become increasingly critical. Understanding these dynamics is essential for ensuring the resilience, accessibility, and ethical use of GNSS technologies in a rapidly evolving global landscape.

CHAPTER 11: THE ROLE OF AI AND MACHINE LEARNING IN GPS SECURITY

Artificial Intelligence (AI) and Machine Learning (ML) are revolutionizing the field of GPS security by addressing challenges such as signal accuracy, resilience to interference, and vulnerability to spoofing and jamming. By leveraging advanced algorithms, AI enhances the robustness of GNSS systems, enabling real-time threat detection and predictive capabilities. This chapter explores how AI and ML are applied to improve GPS security, focusing on enhancing signal accuracy, detecting threats in real time, and implementing predictive analytics for GNSS resilience.

· ·

1. ENHANCING SIGNAL ACCURACY THROUGH AI ALGORITHMS

Challenges in Signal Accuracy

GPS signals are subject to various disruptions, including atmospheric interference, multipath effects, and clock errors. These inaccuracies can degrade navigation and timing services, especially in challenging environments like urban canyons or dense forests.

AI-Powered Solutions

AI algorithms address these challenges by dynamically analyzing and correcting signal disruptions.

1. **Error Correction Models**

   – AI-based models, such as neural networks and decision trees, analyze patterns in GPS data to predict and correct errors.

   – **Applications:**

      • **Atmospheric Delays:** AI models use meteorological data to predict ionospheric and tropospheric delays, adjusting position calculations in real time.

      • **Multipath Interference:** Machine learning

algorithms identify and filter out reflected signals, isolating direct paths from satellites.

2. **Sensor Fusion**

– AI integrates GPS data with inputs from other sensors, such as inertial navigation systems (INS), barometers, and accelerometers.

– **Applications:**

- Enhances accuracy in GPS-denied environments, such as tunnels or underwater regions.

- Reduces reliance on satellite signals in areas with weak coverage.

3. **Signal Prediction**

– AI predicts satellite signal strength and quality based on historical data and real-time observations.

– **Applications:**

- Guides users to areas with better signal conditions.

- Improves performance in dynamic environments like moving vehicles or aircraft.

• •

2. REAL-TIME DETECTION OF JAMMING AND SPOOFING

Threat Landscape

- **Jamming:** Deliberate interference disrupts GPS signals by overwhelming them with noise or stronger signals.

- **Spoofing:** Fake GPS signals mislead receivers, potentially causing significant security breaches.

AI for Jamming Detection

1. **Signal Anomaly Detection**

   - Machine learning models analyze GPS signal parameters, such as power levels, frequency, and timing, to identify anomalies indicative of jamming.

   - **Techniques:**

     - **Unsupervised Learning:** Identifies deviations from normal signal patterns without prior training.

     - **Supervised Learning:** Trains models using labeled datasets of jamming scenarios.

2. **Geospatial Analysis**

   - AI integrates GPS data with spatial mapping to identify

regions experiencing jamming.

- **Applications:**
  - Detects localized jamming zones, such as airports or military installations.
  - Helps authorities trace the source of interference.

3. **Dynamic Frequency Adjustment**

- AI systems dynamically adjust receiver frequencies to avoid jamming bands, maintaining continuous navigation.

• •

AI for Spoofing Detection

1. **Signal Authentication**

- AI algorithms authenticate GPS signals by cross-referencing them with cryptographic keys or known satellite characteristics.

- **Techniques:**
  - **Deep Learning Models:** Detect subtle variations between genuine and spoofed signals.
  - **Multi-GNSS Cross-Validation:** Compares data from multiple GNSS systems (e.g., GPS, Galileo) to detect inconsistencies.

2. **Behavioral Analysis**

- AI monitors the receiver's behavior for abrupt changes in position, speed, or time that may indicate spoofing.

- **Applications:**
  - Flags sudden deviations, such as a ship appearing miles inland.
  - Prevents unauthorized redirection of autonomous vehicles or drones.

3. Real-Time Alerts

–   AI systems generate real-time alerts when spoofing is detected, enabling immediate corrective actions.

• •

3. PREDICTIVE ANALYTICS FOR GNSS RESILIENCE

Understanding Predictive Analytics

Predictive analytics uses AI to anticipate potential disruptions or vulnerabilities, enabling proactive measures to enhance GNSS security.

AI-Driven Predictions

1. **Signal Degradation Forecasting**

   – AI models analyze historical and real-time data to predict periods of poor GPS signal quality.

   – **Applications:**

      • Identifies areas prone to multipath effects or atmospheric delays.

      • Guides users to alternative navigation solutions during disruptions.

2. **Threat Detection and Prevention**

   – Predictive analytics identifies patterns in jamming or spoofing attempts, enabling preemptive responses.

   – **Applications:**

      • Military systems deploy countermeasures in high-risk areas.

      • Civilian users receive warnings about potential

GNSS disruptions.

3. **Satellite Health Monitoring**

–   AI analyzes telemetry data from GNSS satellites to predict maintenance needs or potential failures.

–   **Applications:**

  •   Ensures continuous service by scheduling repairs or replacements.

  •   Prevents loss of coverage in critical regions.

• •

Real-World Applications of Predictive Analytics

1. **Military Operations**

–   AI predicts GNSS vulnerabilities in conflict zones, enhancing navigation security for troops and equipment.

2. **Aviation Safety**

–   Predictive models ensure uninterrupted GNSS service for aircraft by identifying and mitigating threats in advance.

3. **Smart Cities**

–   AI anticipates disruptions to GNSS-reliant systems, such as public transportation or emergency services.

• •

CONCLUSION

AI and machine learning are transforming GPS security by addressing challenges in accuracy, interference detection, and resilience. Through advanced algorithms, real-time monitoring, and predictive analytics, AI enhances the reliability and robustness of GNSS systems in an increasingly complex threat landscape. By integrating these technologies, we can ensure the continued evolution and security of GPS as a critical global utility.

CHAPTER 12: BACKUP SYSTEMS AND ALTERNATIVES

While GPS and other GNSS systems provide unparalleled global navigation and timing capabilities, their vulnerabilities—such as jamming, spoofing, or outages—necessitate the development of backup systems and alternatives. These solutions ensure continuity in critical operations, offering robustness and redundancy. This chapter explores legacy systems like LORAN, modern alternatives such as eLORAN and inertial navigation systems (INS), and hybrid approaches that integrate multiple technologies.

• •

1. OVERVIEW OF ALTERNATIVES

1.1 LORAN and eLORAN

LORAN (Long Range Navigation):

- **History:**

  - Developed during World War II, LORAN was a ground-based radio navigation system providing long-range position fixing.

  - Used extensively for maritime and aviation navigation until phased out in many regions due to GPS adoption.

- **How It Works:**

  - Relies on low-frequency radio signals transmitted by ground-based stations.

  - Receivers measure the time difference between signals from multiple stations to determine position.

eLORAN (Enhanced LORAN):

- **Modern Successor:**

  - eLORAN upgrades LORAN with improved signal accuracy, encryption, and timing capabilities.

- **Features:**

  - Provides positional accuracy of **8–20 meters** and timing accuracy within **50 nanoseconds**.

  - Resistant to GPS jamming and spoofing due to stronger terrestrial signals.

- **Applications:**

  - Maritime navigation, aviation, and national infrastructure resilience.

- **Adoption:**

  - Implemented in countries like the U.S., the U.K., and South Korea as a backup to GNSS.

· ·

1.2 Celestial Navigation

Historical Context:
- Celestial navigation is one of the oldest navigation methods, relying on the positions of celestial bodies such as the sun, moon, stars, and planets.

- Instruments like sextants and almanacs enabled sailors to calculate their position relative to known celestial objects.

Modern Revival:
- **Applications:**

  - Celestial navigation remains a fallback for maritime and military operations when GPS is unavailable.

- **Automated Systems:**

  - Advances in technology allow autonomous systems to use celestial navigation through sensors and AI.

  - Examples: The U.S. Navy's experiments with star trackers for submarines.

· ·

1.3 Inertial Navigation Systems (INS)

Overview:
- INS relies on accelerometers and gyroscopes to measure movement and calculate position relative to a known starting point.

How It Works:

- **Accelerometers:** Measure linear motion.
- **Gyroscopes:** Track angular rotation.
- **Integration:** Combines these measurements to estimate position, velocity, and orientation.

Strengths and Limitations:
- **Advantages:**

  – Immune to external interference like jamming or spoofing.

  – Operates independently without satellite signals.

- **Disadvantages:**

  – Accumulation of errors (drift) over time without external corrections.

- **Applications:**

  – Widely used in aircraft, submarines, and autonomous vehicles as a backup to GPS.

• •

1.4 Hybrid Navigation Systems

Combining Technologies for Resilience:
- Hybrid systems integrate GPS with alternative navigation methods, improving reliability and accuracy.

- **Examples:**

  – **INS + GPS:** Combines the long-term accuracy of GPS with the short-term stability of INS.

  – **Celestial Navigation + INS:** Augments INS with celestial updates to reduce drift.

  – **GPS + eLORAN:** Uses eLORAN as a fallback for GPS disruptions.

Applications:
- **Aviation:** Ensures continuous navigation during GPS

outages.

- **Maritime:** Combines GPS and eLORAN for safe navigation in jamming-prone areas.

- **Autonomous Vehicles:** Integrates GPS, INS, and visual odometry for robust navigation.

• •

2. NATIONAL INITIATIVES FOR GNSS INDEPENDENCE

2.1 Strategic Importance of GNSS Independence

- Reliance on foreign GNSS systems can expose nations to strategic vulnerabilities.

- National GNSS programs enhance security, resilience, and technological self-reliance.

. . . .

2.2 Examples of National Initiatives

United States:
- **eLORAN Deployment:**

    – As part of its GPS modernization efforts, the U.S. is reviving eLORAN as a complementary system.

- **Resilient Navigation Initiatives:**

    – Development of GPS III with advanced anti-jamming capabilities.

European Union:
- **Galileo:**

    – Established to reduce dependence on GPS and ensure European GNSS autonomy.

- **eLORAN Implementation:**

    – Deployed in the U.K. to provide additional resilience for

critical infrastructure.

China:

- **BeiDou:**

  – A global GNSS developed to rival GPS, offering independent navigation capabilities.

- **Integration with Terrestrial Systems:**

  – Combines BeiDou with ground-based systems for enhanced regional accuracy.

Russia:

- **GLONASS:**

  – Focused on maintaining sovereignty over navigation services.

- **Hybrid Solutions:**

  – Integrates GLONASS with regional terrestrial systems for resilience.

India:

- **NavIC:**

  – Regional system focused on high-accuracy services in the Indian subcontinent.

- **Terrestrial Augmentation:**

  – Developing ground-based solutions to complement satellite coverage.

Japan:

- **QZSS:**

  – Augments GPS with regional coverage tailored for the Asia-Pacific region.

- **Resilience Measures:**

  – Incorporates GNSS-independent backup systems for disaster management.

CONCLUSION

While GPS and GNSS systems provide critical navigation services, their vulnerabilities demand robust alternatives and backup systems. Solutions like eLORAN, celestial navigation, and INS offer resilience, while hybrid navigation systems ensure seamless functionality during disruptions. National initiatives for GNSS independence underscore the importance of diversified strategies to safeguard navigation infrastructure. Together, these measures strengthen global navigation reliability in an increasingly interconnected world.

CHAPTER 13: THE FUTURE OF GPS TECHNOLOGY

GPS technology has been a cornerstone of modern navigation, timing, and geolocation systems for decades. However, as the demands for higher accuracy, resilience, and integration with emerging technologies grow, GPS systems must evolve. This chapter explores advancements in atomic clocks and satellite technology, the development of next-generation GPS systems, integration with 5G and IoT, and the potential role of quantum technologies in redefining navigation.

• •

1. ADVANCES IN ATOMIC CLOCK AND SATELLITE TECHNOLOGY

1.1 Atomic Clock Innovations

- **Core Role of Atomic Clocks:**

  - Atomic clocks ensure the precision of GPS signals by providing highly accurate timing, which is critical for position calculation.

- **Next-Generation Atomic Clocks:**

  - **Rubidium Clocks:** Compact and cost-effective, used in current satellites.

  - **Cesium Clocks:** Provide enhanced long-term stability and are increasingly being incorporated into newer systems.

  - **Hydrogen Maser Clocks:** Offer even greater stability and precision, reducing timing errors to improve GPS accuracy.

- **Future Developments:**

  - **Optical Clocks:**

    - Use lasers to measure atomic transitions at optical frequencies.

- Provide 100 times the accuracy of traditional atomic clocks.

- Expected to play a significant role in next-generation GNSS.

· ·

1.2 Satellite Technology Advancements

- **Durability and Efficiency:**

  – New materials and designs extend satellite lifespans beyond the typical 10–15 years.

  – Enhanced solar panel efficiency improves power availability for high-performance systems.

- **Higher Signal Power:**

  – Improved transmission power makes signals more resilient to interference and jamming.

  – Supports multi-frequency operations for better accuracy.

- **Miniaturization:**

  – Smaller, lighter satellites reduce launch costs and allow for faster deployment of constellations.

- **Inter-Satellite Communication:**

  – Enables satellites to share data directly, reducing dependence on ground stations and improving real-time performance.

· ·

2. NEXT-GENERATION SYSTEMS: GPS III AND BEYOND

2.1 GPS III

- **Overview:**

  - The next iteration of the U.S. GPS system, designed to provide greater accuracy, resilience, and capacity.

- **Key Features:**

  - **Higher Accuracy:**

    - Improved atomic clocks and advanced signal structures reduce positional errors.

  - **Stronger Signals:**

    - Increased signal power for enhanced performance in urban and contested environments.

  - **M-Code for Military Use:**

    - Encrypted and resistant to jamming and spoofing.

  - **Civilian L1C Signal:**

    - Interoperable with other GNSS systems, such as Galileo.

- **Deployment:**

  - As of 2024, several GPS III satellites are operational, with additional launches planned.

• •

2.2 Beyond GPS III

- **GPS IIIF:**

  – Incorporates advanced anti-jamming capabilities and new sensors for improved situational awareness.

- **Next-Generation GNSS:**

  – Integration of AI and machine learning for real-time adaptability.

  – Modular satellite designs enable easier upgrades.

- **Global Coordination:**

  – Future GNSS systems may focus on greater interoperability, ensuring seamless use across multiple constellations (e.g., GPS, Galileo, BeiDou).

• •

3. INTEGRATION WITH 5G, IOT, AND EDGE COMPUTING

3.1 5G Integration

- **Enhanced Connectivity:**

    – 5G's low latency and high-speed data transfer complement GPS by enabling faster processing of geolocation data.

- **Applications:**

    – Autonomous vehicles: Real-time positioning with minimal delays.

    – Urban navigation: Seamless integration with smart city infrastructure.

- **Network-Based Augmentation:**

    – 5G networks can provide localized corrections to improve GPS accuracy in dense urban environments.

• •

3.2 IoT Integration

- **Geolocation for IoT Devices:**

    – IoT devices increasingly rely on GPS for tracking, monitoring, and automation.

- **Applications:**

- **Supply Chain Management:** Real-time tracking of goods.

- **Smart Cities:** Coordination of services like waste management and public transportation.

- **Wearables:** Enhanced fitness and health tracking.

• **Challenges and Solutions:**

- GPS power consumption is a concern for IoT devices; advances in low-power GNSS chips are addressing this issue.

• •

3.3 Edge Computing

• **Overview:**

- Edge computing processes data closer to the source, reducing latency and bandwidth usage.

• **Role in GPS:**

- Localized processing of GPS data enhances accuracy and speed.

- Reduces reliance on centralized cloud infrastructure for time-critical applications.

• •

4. POTENTIAL ROLE OF QUANTUM TECHNOLOGIES IN NAVIGATION

4.1 Quantum Clocks

- **Next-Level Precision:**

    – Quantum clocks, based on quantum entanglement and superposition, are expected to provide unprecedented timing accuracy.

    – Reduces positional errors to mere millimeters.

- **Integration with GNSS:**

    – Quantum clocks may replace current atomic clocks in satellites, revolutionizing timing and navigation capabilities.

. . . .

4.2 Quantum Sensors

- **Overview:**

    – Quantum sensors use the principles of quantum mechanics to detect minute changes in gravitational fields and inertial forces.

- **Applications:**

    – **Quantum Gravimeters:**

- Measure local gravitational anomalies to calculate position without satellite signals.
  - **Quantum Inertial Navigation:**
    - Provides GPS-independent navigation for submarines, aircraft, and spacecraft.

• •

4.3 Quantum Cryptography

- **Enhanced Security:**
  - Quantum cryptography could provide unbreakable encryption for GNSS signals, preventing spoofing and tampering.
- **Applications:**
  - Critical for military and national security operations.

• •

CONCLUSION

The future of GPS technology is marked by rapid advancements in satellite design, clock precision, and integration with emerging technologies. Next-generation systems like GPS III, combined with 5G, IoT, and edge computing, promise unprecedented accuracy and resilience. The integration of quantum technologies could revolutionize navigation, offering unparalleled precision and security. As GPS continues to evolve, it will remain a cornerstone of global infrastructure, driving innovation across industries and improving connectivity worldwide.

CHAPTER 14: ETHICAL AND GEOPOLITICAL CHALLENGES

The widespread reliance on GPS and other GNSS systems raises significant ethical and geopolitical challenges. As these systems play a critical role in global infrastructure, balancing open access with security, addressing the ethical implications of signal manipulation, and navigating geopolitical tensions are essential. This chapter explores these issues in depth, using real-world case studies to illustrate the complexities of GPS in a global context.

• •

1. BALANCING OPEN ACCESS WITH SECURITY NEEDS

1.1 Importance of Open Access

- **Global Utility:**

    - GPS is a public good, providing essential services for navigation, timing, and geolocation worldwide.

    - Open access enables widespread adoption across civilian, commercial, and scientific domains.

- **Economic Benefits:**

    - Industries such as transportation, agriculture, and telecommunications rely on freely available GPS signals.

    - A 2019 study estimated GPS contributes over $1.4 trillion annually to the U.S. economy alone.

1.2 Security Concerns

- **National Security Risks:**

    - Open GPS signals can be exploited by adversaries for military operations, reconnaissance, and cyberattacks.

    - Civilian systems are vulnerable to jamming, spoofing, and other forms of interference.

- **Encryption and Restricted Access:**

    - Military GNSS signals, such as GPS's M-code, are encrypted to prevent unauthorized use.

–   Civilian systems must strike a balance between providing open access and protecting critical infrastructure.

1.3 Balancing Acts

- **Selective Availability (Historical Example):**

–   Until 2000, the U.S. degraded GPS accuracy for civilian users through selective availability to maintain military superiority.

–   The policy was discontinued to promote innovation and economic growth but highlights the tension between access and security.

- **Regional Alternatives:**

–   Countries like China (BeiDou) and Russia (GLONASS) have developed independent GNSS systems to ensure national security and reduce reliance on GPS.

• •

2. ETHICAL CONSIDERATIONS OF GPS MANIPULATION

2.1 GPS Manipulation Defined

- Manipulation involves deliberately altering GPS signals to achieve strategic, military, or political objectives. Techniques include:

  - **Jamming:** Blocking GPS signals to disrupt navigation.

  - **Spoofing:** Transmitting fake signals to mislead receivers.

2.2 Ethical Implications

- **Civilian Harm:**

  - GPS manipulation can disrupt emergency services, aviation, and maritime navigation, endangering lives.

  - Example: Jamming near airports can interfere with aircraft landings.

- **Economic Impact:**

  - Spoofing or jamming can disrupt logistics and supply chains, leading to significant financial losses.

- **Surveillance and Privacy Concerns:**

  - GNSS signals can be exploited for mass surveillance, raising ethical questions about the balance between security and individual privacy.

- **Dual-Use Technology:**

    – While GNSS serves civilian needs, its military applications can escalate conflicts and trigger arms races.

2.3 Legal and Regulatory Frameworks

- **International Agreements:**

    – The Outer Space Treaty (1967) prohibits harmful interference with space-based systems but lacks enforcement mechanisms.

- **National Laws:**

    – Countries impose penalties for unauthorized GPS manipulation, yet enforcement is challenging in cross-border scenarios.

• •

3. CASE STUDIES: GEOPOLITICAL CONFLICTS INVOLVING GNSS SYSTEMS

3.1 North Korea's Jamming Campaigns

- **Incident:**

  - Between 2010 and 2016, North Korea repeatedly jammed GPS signals near its border, targeting South Korea.

- **Impact:**

  - Disrupted aviation, shipping, and telecommunications.

  - Highlighted the vulnerability of GNSS-dependent infrastructure to state-sponsored interference.

- **Response:**

  - South Korea enhanced its GNSS resilience by deploying eLORAN and monitoring systems.

• •

3.2 Black Sea Spoofing Incident

- **Incident:**

- In 2017, ships in the Black Sea reported GPS anomalies, showing positions miles inland.

- **Suspected Actors:**

  - Russia was suspected of using GNSS spoofing as part of its electronic warfare strategy.

- **Impact:**

  - Raised concerns about the use of GPS manipulation in geopolitical conflicts.

  - Prompted NATO to develop counter-spoofing measures.

• •

3.3 Iranian Capture of U.S. Drone

- **Incident:**

  - In 2011, Iran reportedly spoofed GPS signals to mislead a U.S. RQ-170 Sentinel drone into landing in Iranian territory.

- **Implications:**

  - Demonstrated the strategic use of GPS manipulation in modern warfare.

  - Accelerated efforts to secure military GNSS signals against spoofing.

• •

3.4 China's BeiDou Diplomacy

- **Incident:**

  - China has promoted its BeiDou system as an alternative to GPS, particularly among developing nations.

- **Motivation:**

  - Reduces reliance on U.S.-controlled GPS, increasing China's geopolitical influence.

- **Impact:**

–   BeiDou integration in Africa and Asia fosters economic ties but raises concerns about surveillance and data sovereignty.

• •

4. FUTURE CHALLENGES AND SOLUTIONS

4.1 Rising GNSS Militarization

- Increased use of GNSS for military applications, such as precision-guided munitions, escalates global tensions.

- Nations are investing in countermeasures like anti-satellite weapons (ASAT) and electronic warfare systems.

4.2 Need for International Cooperation

- Strengthening international agreements to prevent GNSS manipulation is essential.

- Collaborative initiatives, like GNSS interoperability, can reduce geopolitical tensions and enhance global resilience.

4.3 Ethical Use of Emerging Technologies

- As AI and quantum technologies integrate with GNSS, ethical guidelines are needed to ensure responsible development and deployment.

- Balancing innovation with security and privacy will remain a central challenge.

CONCLUSION

The ethical and geopolitical challenges of GPS technology highlight the complex interplay between global accessibility, national security, and ethical responsibility. While GPS and GNSS systems provide immense benefits, their vulnerabilities and potential misuse demand robust regulatory frameworks, international cooperation, and ethical oversight. Addressing these challenges ensures the continued reliability and equitable use of GPS in an interconnected world.

CHAPTER 15: SETTING UP A GPS SECURITY LAB

Establishing a GPS security lab provides researchers, technologists, and security professionals with a controlled environment to study GPS vulnerabilities, develop mitigation techniques, and test new applications. This chapter provides a comprehensive guide to setting up a GPS security lab, focusing on equipment requirements, legal considerations, and best practices for maintaining a controlled and ethical testing environment.

• •

1. EQUIPMENT REQUIREMENTS

A well-equipped GPS security lab requires specialized hardware and software to simulate, analyze, and mitigate GPS-related scenarios. Below are the essential components:

1.1 Signal Analyzers

- **Purpose:**

  - To monitor and analyze GPS signal characteristics, such as frequency, power, and timing.

- **Key Features:**

  - High-frequency range to capture GNSS signals (e.g., L1, L2, L5 bands).

  - Real-time spectrum analysis for detecting interference or jamming.

- **Examples:**

  - Rohde & Schwarz FSW Signal Analyzer.

  - Keysight N9030A PXA Signal Analyzer.

. . . .

1.2 GPS Receivers

- **Purpose:**

  - To receive and process GPS signals, providing data for analysis.

- **Types:**

–   **Consumer-Grade Receivers:** Test general GPS functionality (e.g., Garmin, smartphone GNSS chips).

–   **Survey-Grade Receivers:** Provide high-accuracy data for professional applications (e.g., Trimble, Leica).

–   **Military-Grade Receivers:** (Restricted access) Test resilience to jamming and spoofing in defense applications.

1.3 Antennas

- **Purpose:**

  –   To capture GPS signals for analysis and testing.

- **Types:**

  –   **High-Gain Antennas:** Improve signal reception in low-strength environments.

  –   **Multi-Band Antennas:** Support multiple GNSS systems (e.g., GPS, Galileo, GLONASS, BeiDou).

  –   **Choke-Ring Antennas:** Minimize multipath interference for precise measurements.

1.4 GPS Signal Generators

- **Purpose:**

  –   To simulate GPS signals for testing receivers and analyzing vulnerabilities.

- **Key Features:**

  –   Multi-frequency support for generating complex scenarios.

  –   Capability to simulate spoofing or jamming for controlled experiments.

- **Examples:**

- Spirent GSS7000 Series GNSS Simulator.

- Rohde & Schwarz SMW200A Vector Signal Generator.

. . . .

1.5 Software and Tools

- **Purpose:**

  - To analyze data, simulate scenarios, and automate testing processes.

- **Recommended Tools:**

  - **GNURadio:** Open-source software for processing and analyzing GNSS signals.

  - **MATLAB:** Advanced platform for signal modeling and algorithm development.

  - **GPS Toolkit (GPSTk):** Open-source library for GPS data processing.

  - **Skydel SDX:** Software-defined GNSS simulation tool.

. . . .

2. LEGAL CONSIDERATIONS FOR TESTING

Conducting GPS security experiments involves potential risks to public safety and requires adherence to legal and ethical standards.

2.1 Regulatory Compliance

- **National Regulations:**

    – Many countries regulate GPS testing to prevent interference with civilian and military operations.

    – Example: The U.S. Federal Communications Commission (FCC) prohibits unauthorized jamming or spoofing.

- **International Standards:**

    – Comply with ITU guidelines for radio frequency use and interference prevention.

2.2 Testing Licenses and Permits

- **Obtaining Authorization:**

    – Apply for testing permits from relevant authorities (e.g., FCC in the U.S., Ofcom in the U.K.).

    – Outline the scope of experiments, equipment used, and safety measures.

- **Restricted Environments:**

- Perform experiments in shielded or remote areas to prevent unintentional interference with public GPS services.

2.3 Ethical Considerations

- **Minimize Risks:**

  - Ensure that experiments do not endanger lives, disrupt infrastructure, or violate privacy.

- **Transparency:**

  - Clearly document testing procedures, objectives, and outcomes.

· ·

3. BEST PRACTICES FOR CREATING A CONTROLLED ENVIRONMENT

A controlled environment ensures the safety, reliability, and reproducibility of GPS security experiments.

3.1 Shielded Test Environments

- **Anechoic Chambers:**

    - Shielded rooms that block external radio frequency interference, ideal for precise signal testing.

- **Faraday Cages:**

    - Enclosures made of conductive materials to isolate testing equipment from external signals.

3.2 Signal Isolation

- **Directional Antennas:**

    - Use antennas that focus on specific signals to reduce interference.

- **Frequency Filters:**

    - Employ filters to isolate GNSS frequencies and block extraneous noise.

3.3 Experimental Setup

- **Simulating Scenarios:**

- – Use signal generators to replicate real-world conditions, such as jamming, spoofing, or multipath interference.

- **Monitoring Systems:**

- – Continuously monitor signal strength, accuracy, and timing during experiments.

3.4 Data Collection and Analysis

- **Logging Tools:**

- – Use data loggers to record signal parameters, receiver behavior, and environmental conditions.

- **Automation:**

- – Automate repetitive tasks using scripts and software tools to improve efficiency.

• •

3.5 Safety Protocols

- **Emergency Shutdown:**

- – Implement fail-safe mechanisms to immediately terminate experiments if unintended interference occurs.

- **Periodic Calibration:**

- – Regularly calibrate equipment to ensure accurate and reliable results.

• •

CONCLUSION

Establishing a GPS security lab is essential for advancing GNSS research, developing robust defense mechanisms, and testing innovative applications. By procuring the right equipment, adhering to legal and ethical standards, and maintaining a controlled testing environment, researchers can safely explore the complexities of GPS security. This structured approach ensures the lab's contributions to the resilience and reliability of global navigation systems.

CHAPTER 16: PRACTICAL TUTORIALS

Practical experimentation is essential for understanding GPS vulnerabilities and testing solutions to counter threats such as jamming and spoofing. This chapter provides step-by-step tutorials for simulating jamming scenarios in a controlled and legal setup, testing anti-jamming devices, and detecting and countering spoofed signals using adaptive antennas and cryptographic tools.

• •

1. SIMULATING JAMMING SCENARIOS IN A SAFE AND LEGAL SETUP

Simulating jamming scenarios enables researchers to study interference effects and develop countermeasures. This tutorial focuses on conducting such experiments in a safe, controlled, and legally compliant manner.

1.1 Prerequisites

- **Equipment Required:**

  - GPS receiver with logging capabilities.

  - Signal generator or low-power jammer (for controlled use only).

  - Signal analyzer to monitor interference effects.

  - Shielded test environment (e.g., Faraday cage or anechoic chamber).

- **Legal Compliance:**

  - Obtain necessary permits from regulatory authorities.

  - Ensure the test environment prevents interference with external GNSS signals.

1.2 Steps

1. **Set Up the Environment:**

- Position the GPS receiver inside a shielded enclosure to ensure isolation.

- Connect the signal generator or jammer to an antenna.

2. **Configure the Signal Generator:**

- Set the generator to emit signals at GPS L1 frequency (1575.42 MHz) or the relevant GNSS frequency.

- Adjust power levels to simulate various jamming intensities (low to high).

3. **Run the Experiment:**

- Begin with no jamming and record baseline GPS performance.

- Gradually increase jamming power and observe its effect on receiver accuracy, signal strength, and timing.

4. **Analyze the Results:**

- Use the signal analyzer to monitor interference patterns.

- Compare receiver performance under different jamming intensities.

5. **Document Findings:**

- Record all observations and identify thresholds where the receiver begins to fail.

1.3 Applications

• Evaluate the resilience of commercial and military-grade receivers.

• Test the effectiveness of anti-jamming techniques.

• •

2. TESTING ANTI-JAMMING DEVICES WITH ADAPTIVE ANTENNAS

Adaptive antennas, such as phased arrays, dynamically adjust their reception patterns to reject interference. This tutorial demonstrates how to evaluate their performance in countering jamming.

2.1 Prerequisites

- **Equipment Required:**
    - Adaptive antenna system.
    - GPS receiver compatible with adaptive antennas.
    - Signal generator or jammer.
    - Spectrum analyzer.
- **Test Environment:**
    - Shielded lab setup to prevent unintended interference.

2.2 Steps

1. **Prepare the Setup:**
    - Connect the adaptive antenna to the GPS receiver.
    - Place the receiver in a shielded environment.
2. **Introduce Jamming:**

– Use a signal generator to emit jamming signals at various frequencies and power levels.

3. **Monitor Antenna Behavior:**

– Observe how the adaptive antenna adjusts its reception pattern to nullify jamming signals.

– Use the spectrum analyzer to verify the reduction in interference.

4. **Evaluate Performance:**

– Record the receiver's accuracy, signal-to-noise ratio (SNR), and ability to maintain a lock on satellite signals under jamming conditions.

2.3 Applications

• Validate the effectiveness of adaptive antennas in real-world scenarios.

• Develop guidelines for deploying adaptive systems in critical infrastructure.

• •

3. DETECTING AND COUNTERING SPOOFED SIGNALS

Spoofing attacks deceive GPS receivers by broadcasting fake signals. Detecting and mitigating spoofing requires understanding signal patterns and using cryptographic tools.

* *

3.1 Recognizing Spoofing Patterns

Steps to Detect Spoofing

1. **Monitor Signal Behavior:**

   – Use a signal analyzer to identify anomalies in signal strength, timing, and frequency.

   – Sudden increases in signal strength or multiple signals from a single satellite are potential indicators of spoofing.

2. **Analyze Receiver Behavior:**

   – Check for unexpected jumps in position or time.

   – Sudden and implausible movements (e.g., being relocated miles away) suggest spoofing.

3. **Cross-Validate Signals:**

   – Use data from multiple GNSS constellations (e.g., Galileo, GLONASS) to verify consistency.

   – Compare the receiver's output with known satellite

positions.

Applications

- Real-time detection of spoofing attacks in civilian and military systems.

- Enhance the reliability of navigation for critical operations.

• •

3.2 Using Cryptographic Tools

Cryptographic authentication ensures the integrity and authenticity of GPS signals, making spoofing significantly more challenging.

Steps to Implement Cryptographic Verification

1. **Obtain Cryptographic Keys:**

   – Access public key infrastructure (PKI) data for authenticated GNSS signals.

   – Encrypted signals like GPS M-code or Galileo PRS require authorized access.

2. **Enable Cryptographic Modules:**

   – Configure the GPS receiver to validate signals using cryptographic keys.

   – Ensure the receiver can reject unauthenticated signals.

3. **Test Signal Authentication:**

   – Simulate spoofed signals without cryptographic signatures.

   – Verify that the receiver rejects these signals and maintains accurate navigation.

4. **Monitor Authentication Logs:**

   – Use logging tools to track signal validation processes.

   – Identify and analyze instances of rejected spoofed signals.

Applications

- Secure critical systems, such as financial networks, power grids, and military operations.

- Develop robust receivers for applications requiring high integrity.

• •

4. PRACTICAL SCENARIOS AND INSIGHTS

- **Military Testing:** Simulate GPS spoofing to evaluate the resilience of encrypted GNSS signals in combat environments.

- **Commercial Use Cases:** Test anti-jamming and spoofing defenses in transportation, aviation, and autonomous systems.

- **Research and Development:** Develop algorithms for detecting interference and authenticating signals in real-time.

• •

CONCLUSION

Simulating and countering GPS vulnerabilities in a controlled lab environment provides valuable insights into the robustness of navigation systems. By leveraging practical tools, adaptive technologies, and cryptographic authentication, researchers and professionals can enhance the security and reliability of GPS systems in an evolving threat landscape. These tutorials serve as a foundation for advancing GPS resilience in critical applications.

CHAPTER 17: PROTECTING CRITICAL INFRASTRUCTURE

Critical infrastructure sectors, such as aviation, shipping, and energy, rely heavily on GPS for navigation, timing, and operational efficiency. However, these dependencies expose them to vulnerabilities, including jamming, spoofing, and outages. Protecting critical infrastructure involves implementing security measures, redundancy plans, and robust maintenance protocols. This chapter outlines the steps required to secure GPS-reliant systems, integrate alternative navigation solutions, and establish regular monitoring practices.

• •

1. SECURING GPS-RELIANT SYSTEMS

1.1 Aviation

- **Role of GPS:**

  1. Provides navigation for en-route flights, approach, and landing.

  2. Synchronizes timing for air traffic control systems and communication networks.

- **Steps to Secure GPS in Aviation:**

  1. **Deploy Anti-Jamming and Anti-Spoofing Technologies:**

     - Equip aircraft with adaptive antennas to counter jamming.

     - Use authenticated signals, such as SBAS (Satellite-Based Augmentation System), for secure navigation.

  2. **Integrate Backup Systems:**

     - Pair GPS with Inertial Navigation Systems (INS) to ensure navigation continuity during GPS disruptions.

  3. **Training and Simulation:**

     - Conduct regular training for pilots and air traffic controllers to handle GPS outages.

     - Simulate interference scenarios to test the

resilience of systems.

4. **Establish Geo-Fencing:**

- Use geo-fencing tools to detect unauthorized GNSS interference around airports.

1.2 Shipping

- **Role of GPS:**

  1. Guides vessels through oceans and ports, enhances collision avoidance, and facilitates efficient routing.

  2. Tracks goods and optimizes supply chain logistics.

- **Steps to Secure GPS in Shipping:**

  1. **Implement GNSS Redundancy:**

     - Use multi-constellation GNSS receivers (e.g., GPS, Galileo, BeiDou) to reduce reliance on a single system.

     - Integrate eLORAN as a terrestrial navigation backup.

  2. **Deploy Automated Monitoring Systems:**

     - Install systems that continuously check signal integrity and alert for anomalies.

  3. **Enhance Security at Ports:**

     - Monitor GNSS signals within port areas to detect and mitigate jamming or spoofing.

1.3 Energy

- **Role of GPS:**

  1. Synchronizes power grids and telecommunication networks.

  2. Facilitates timing for financial transactions and operational control.

- **Steps to Secure GPS in Energy:**

  1. **Hardened Timing Systems:**

     - Deploy precision timing technologies like atomic clocks or quantum clocks as alternatives to GPS timing.

  2. **Resilient Communication Networks:**

     - Use encrypted communication protocols to prevent signal manipulation.

  3. **Cybersecurity Integration:**

     - Incorporate GPS signal monitoring into broader cybersecurity frameworks to detect anomalies.

• •

2. REDUNDANCY PLANNING AND INTEGRATION OF ALTERNATIVE NAVIGATION SYSTEMS

2.1 Importance of Redundancy

- Redundancy ensures uninterrupted operations during GPS outages or disruptions.

- Mitigates the risks of economic loss, safety hazards, and service downtime.

· ·

2.2 Alternative Navigation Systems

1. **Inertial Navigation Systems (INS):**

   - Uses accelerometers and gyroscopes to calculate position independently of external signals.

   - Strengths:

     - Immune to jamming and spoofing.

     - Provides short-term navigation accuracy during GPS outages.

2. **eLORAN:**

– A terrestrial navigation system providing complementary coverage to GNSS.

– Advantages:

  • Stronger signals resistant to jamming.

  • Accurate timing and navigation over long distances.

3. **Celestial Navigation:**

– Automated systems use star trackers to calculate position based on celestial objects.

– Applications:

  • Ideal for maritime and aerospace industries as a backup during GPS disruptions.

4. **Visual Navigation and Sensor Fusion:**

– Combines optical systems, cameras, and radar for position estimation.

– Used in autonomous vehicles and drones to enhance navigation reliability.

• •

2.3 Integrating Redundancy Plans

• **Multi-GNSS Receivers:**

– Enable devices to access signals from multiple constellations (e.g., GPS, Galileo, GLONASS, BeiDou).

– Enhance accuracy and resilience against localized GNSS disruptions.

• **Hybrid Navigation Systems:**

– Combine GPS with INS and other systems for robust performance.

- Ensure seamless switching between systems during disruptions.

· ·

3. REGULAR MAINTENANCE AND MONITORING PROTOCOLS

3.1 Importance of Maintenance

- Proactive maintenance ensures that GPS-dependent systems remain functional and secure.

- Prevents issues from escalating into critical failures.

· ·

3.2 Establishing Monitoring Systems

1. **Continuous Signal Monitoring:**

   - Deploy tools to measure signal strength, timing, and accuracy.

   - Use anomaly detection algorithms to identify interference, spoofing, or degradation.

2. **Real-Time Alerts:**

   - Implement automated alert systems to notify operators of potential threats.

   - Integrate alerts with response protocols to minimize disruption.

· ·

3.3 Regular Testing

1. **Interference Simulations:**

   – Conduct periodic simulations of jamming or spoofing scenarios.

   – Evaluate the effectiveness of mitigation measures.

2. **Performance Audits:**

   – Test GPS receivers and antennas for compliance with industry standards.

   – Ensure software and firmware are up to date with the latest security patches.

• •

3.4 Maintenance Best Practices

1. **Equipment Calibration:**

   – Regularly calibrate GPS receivers, antennas, and signal analyzers for optimal performance.

2. **Data Backup:**

   – Maintain backups of navigation and timing data to ensure continuity during outages.

3. **Collaboration with GNSS Operators:**

   – Share feedback with GNSS providers to improve signal integrity and system updates.

• •

CONCLUSION

Securing GPS-reliant systems is a critical priority for safeguarding aviation, shipping, and energy infrastructure. By implementing redundancy planning, integrating alternative navigation systems, and establishing regular maintenance protocols, organizations can enhance the resilience and reliability of their operations. These measures ensure the continued functionality of critical infrastructure in an increasingly interconnected and vulnerable world.

CONCLUSION

GPS technology has transformed modern life, becoming a cornerstone of global infrastructure. Its applications span navigation, communication, energy synchronization, and defense, underscoring its indispensable role in the interconnected world. However, GPS is not invincible. Its vulnerabilities—ranging from jamming and spoofing to geopolitical manipulation—pose significant risks to safety, security, and economic stability. This conclusion recaps GPS's transformative impact, highlights the importance of safeguarding it, and calls for a balanced approach to innovation, security, and accessibility.

• •

RECAP OF GPS'S TRANSFORMATIVE ROLE AND VULNERABILITIES

Transformative Role of GPS

- **Global Navigation and Timing:**

    - GPS powers navigation for aviation, maritime, and land transportation, ensuring efficient and safe travel.

    - Its precise timing capabilities synchronize power grids, financial transactions, and communication networks.

- **Technological and Economic Enabler:**

    - GPS has driven innovation in autonomous systems, IoT devices, and geospatial technologies.

    - Its economic impact is profound, with industries worldwide benefiting from its accessibility and accuracy.

- **Military and Security Applications:**

    - GPS enhances defense capabilities through precision-guided munitions, troop tracking, and secure communications.

    - It underpins national security strategies, enabling real-time situational awareness.

Vulnerabilities

- **Jamming and Spoofing:**

  – These threats can disrupt navigation, create false positions, and compromise critical operations.

- **Dependence and Redundancy:**

  – Overreliance on GPS without adequate backups leaves systems exposed to failures or attacks.

- **Geopolitical Exploitation:**

  – GNSS systems are increasingly weaponized in global conflicts, creating challenges for neutrality and security.

- **Environmental and Technical Limitations:**

  – Urban canyons, dense foliage, and atmospheric interference can degrade GPS accuracy.

• •

THE IMPORTANCE OF SAFEGUARDING GPS SYSTEMS IN AN INTERCONNECTED WORLD

Protecting Critical Infrastructure

- **Aviation, Shipping, and Energy:**

    – GPS is integral to these sectors, and any disruption could have catastrophic consequences.

    – Robust defenses and alternative systems ensure operational continuity and safety.

Securing Global Trust

- GPS is a shared global resource, and its reliability fosters international cooperation and economic growth.

- Safeguarding it requires collaboration across nations, industries, and regulatory bodies to mitigate risks and enhance resilience.

Innovation with Resilience

- Advancements in AI, machine learning, and quantum technologies offer opportunities to enhance GPS capabilities and security.

- These innovations must be paired with strategies to address emerging threats and vulnerabilities.

· ·

CALL TO ACTION: BALANCING INNOVATION, SECURITY, AND ACCESSIBILITY

1. **Foster Innovation:**

   – Encourage the development of next-generation systems like GPS III, Galileo, and BeiDou to improve accuracy, reliability, and resilience.

   – Invest in research on quantum technologies and hybrid navigation systems for future-proof solutions.

2. **Strengthen Security:**

   – Implement anti-jamming and anti-spoofing technologies in critical infrastructure.

   – Promote cryptographic authentication for signal validation and enhance GNSS resilience through redundancy.

3. **Ensure Accessibility:**

   – Balance security with open access to ensure GPS remains a global public good.

   – Advocate for international cooperation to prevent

misuse and foster trust in GNSS systems.

4. **Build Awareness:**

–   Educate policymakers, industries, and the public about GPS vulnerabilities and their potential impacts.

–   Promote the adoption of best practices and regular testing to strengthen system defenses.

• •

FINAL THOUGHTS

The journey of GPS technology reflects humanity's ingenuity in navigating an increasingly complex world. As we move forward, the challenge lies in preserving its transformative potential while safeguarding against its vulnerabilities. By balancing innovation, security, and accessibility, we can ensure that GPS continues to empower individuals, industries, and nations—building a safer, more connected future.

GLOSSARY OF TECHNICAL TERMS

This glossary provides concise definitions of the technical terms used throughout the book, offering clarity on key concepts and technologies associated with GPS and GNSS systems.

• •

Adaptive Antennas

- **Definition:** Antennas that dynamically adjust their reception patterns to minimize interference, such as jamming or multipath signals.

- **Application:** Used in aviation, defense, and critical infrastructure to ensure signal reliability.

Anechoic Chamber

- **Definition:** A shielded room designed to block electromagnetic interference and reflect radio waves, creating a controlled environment for testing.

- **Application:** Commonly used for GPS signal experiments.

Atomic Clock

- **Definition:** A highly accurate clock that measures time based on the vibrations of atoms, typically cesium or rubidium.

- **Application:** Provides precise timing for GPS satellites, ensuring accurate positioning.

Augmentation System

- **Definition:** A system that improves GNSS accuracy

by providing correction data, often through ground-based or satellite networks.

- **Examples:** WAAS (Wide Area Augmentation System), EGNOS (European Geostationary Navigation Overlay Service).

Beamforming
- **Definition:** A signal processing technique that directs radio wave signals to specific locations, reducing interference.

- **Application:** Used in adaptive antennas to enhance GPS signal reception.

Celestial Navigation
- **Definition:** A traditional navigation method that uses observations of celestial objects (e.g., stars, the sun) to determine position.

- **Modern Application:** Automated celestial navigation systems for maritime and aerospace industries.

Choke-Ring Antenna
- **Definition:** A specialized GPS antenna design that minimizes multipath interference by using concentric metal rings.

- **Application:** Frequently used in high-precision applications like surveying.

Cryptographic Authentication
- **Definition:** The use of cryptographic keys to verify the authenticity of GPS signals, preventing spoofing attacks.

- **Example:** GPS M-code for military applications.

Differential GPS (DGPS)
- **Definition:** A technique that uses ground reference stations to provide correction data, improving GPS accuracy.

- **Accuracy:** Typically enhances positioning to within 1–3 meters.

Ephemeris Data

- **Definition:** Information about the orbital positions of GPS satellites, used by receivers to calculate precise locations.

- **Application:** Broadcast regularly by satellites as part of navigation messages.

eLORAN (Enhanced Long-Range Navigation)

- **Definition:** A terrestrial navigation system that provides a robust backup to GNSS, offering strong signals and high accuracy.

- **Application:** Used for critical infrastructure and maritime navigation.

Frequency Bands (L1, L2, L5)

- **Definition:** The specific radio frequencies used by GPS satellites to transmit signals.

- **Examples:**

  - **L1 (1575.42 MHz):** Used for civilian navigation.

  - **L2 (1227.60 MHz):** Used for military and advanced civilian applications.

  - **L5 (1176.45 MHz):** Provides high-accuracy and robust signals for safety-critical applications.

Geofencing

- **Definition:** A virtual boundary defined using GPS or GNSS, often used to restrict or monitor movement within a specific area.

- **Application:** Used in drones, autonomous vehicles, and security systems.

GNSS (Global Navigation Satellite System)

- **Definition:** A general term for satellite-based navigation systems, including GPS, Galileo, GLONASS, and BeiDou.

- **Coverage:** Provides global or regional positioning services.

Hybrid Navigation System

- **Definition:** A system that combines multiple navigation methods (e.g., GPS, INS, eLORAN) to improve resilience and accuracy.

- **Application:** Used in autonomous vehicles, aviation, and maritime navigation.

Inertial Navigation System (INS)

- **Definition:** A navigation system that calculates position based on motion sensors (accelerometers and gyroscopes) without external signals.

- **Strength:** Immune to jamming and spoofing.

Jamming

- **Definition:** The deliberate interference with GPS signals by transmitting noise or stronger signals on the same frequency.

- **Impact:** Can disrupt navigation, timing, and positioning.

Keplerian Orbits

- **Definition:** The elliptical paths followed by satellites, described by Kepler's laws of planetary motion.

- **Application:** Used to calculate satellite positions for GNSS systems.

Multipath Effect

- **Definition:** Signal distortion caused by GPS signals reflecting off surfaces like buildings or water before reaching the receiver.

- **Impact:** Reduces positioning accuracy, especially in urban environments.

Phased Array

- **Definition:** An antenna system with multiple elements that can change the direction of its beam electronically.

- **Application:** Used in advanced anti-jamming technologies.

Pseudorandom Noise (PRN) Code
- **Definition:** A unique digital code transmitted by each satellite to identify its signals and calculate distances.
- **Application:** Essential for determining position in GPS receivers.

Real-Time Kinematic (RTK) Positioning
- **Definition:** A technique that enhances GPS accuracy to the centimeter level using real-time correction data from a base station.
- **Application:** Used in surveying, precision farming, and construction.

Relativity (GPS Context)
- **Definition:** The theory of relativity accounts for time dilation effects on GPS satellites due to their high speeds (special relativity) and weaker gravity (general relativity).
- **Impact:** Requires adjustments to satellite clocks to ensure accurate timing.

Selective Availability
- **Definition:** A past U.S. policy of intentionally degrading GPS accuracy for civilian users to maintain military advantage.
- **Status:** Discontinued in 2000.

Signal Spoofing
- **Definition:** Broadcasting fake GPS signals to mislead receivers into calculating incorrect positions or times.
- **Impact:** Threatens navigation safety and data integrity.

Spread-Spectrum Technology
- **Definition:** A method of transmitting signals over a wide frequency range to make them more resilient to interference.
- **Application:** Used in GPS signals to prevent jamming.

WAAS (Wide Area Augmentation System)

- **Definition:** A U.S.-based satellite augmentation system that improves GPS accuracy and reliability for aviation and other applications.

- **Accuracy:** Reduces errors to less than 1 meter.

• •

CONCLUSION

This glossary provides a foundational understanding of the technical terms associated with GPS and GNSS systems. By familiarizing yourself with these definitions, you can better appreciate the complexity and innovation behind modern navigation technologies.

MATHEMATICAL FOUNDATIONS OF GPS

The operation of GPS relies on a combination of geometry, algebra, and advanced physics. Understanding the mathematical foundations requires diving into trilateration, signal timing, error correction, and relativistic adjustments. This chapter explores these principles in depth for advanced readers.

• •

1. TRILATERATION: DETERMINING POSITION

1.1 Basic Concept

- Trilateration calculates a receiver's position by measuring its distance from multiple satellites.

- For 2D space, three satellites are required; for 3D space, four are necessary to account for clock errors.

1.2 Mathematical Representation

1. **Distance Equation:**

   – The distance between a satellite and a receiver is given by: $d_i = c \cdot (t_r - t_{si})$ where:

     - d_i: Distance to the i-th satellite.

     - c: Speed of light (3×10^8 m/s).

     - t_r: Receiver clock time.

     - t_{si}: Satellite clock time.

2. **Position Equation:**

   – Each satellite provides a sphere of possible positions:

$$(x - x_i)^2 + (y - y_i)^2 + (z - z_i)^2 = d_i^2$$ where:

- (x,y,z): Receiver's coordinates.

- (x_i, y_i, z_i): Satellite's coordinates.

- d_i: Distance to the satellite.

3. **System of Equations:**

$$(x - x_1)^2 + (y - y_1)^2 + (z - z_1)^2 = d_1^2,$$
$$(x - x_2)^2 + (y - y_2)^2 + (z - z_2)^2 = d_2^2,$$
$$\vdots$$

– For n satellites: $(x - x_n)^2 + (y - y_n)^2 + (z - z_n)^2 = d_n^2.$

4. **Solution:**

– The equations are solved using numerical methods (e.g., least squares) to find (x,y,z).

. . . .

2. SIGNAL TIMING AND CLOCK BIAS

2.1 Correcting for Clock Bias

- GPS receivers use quartz clocks, which are less accurate than satellite atomic clocks, introducing bias b_r.

- Adjusted distance equation: $d_i = c \cdot [(t_r - b_r) - t_{si}]$.

- The receiver's position and clock bias are calculated simultaneously by solving for (x, y, z, b_r).

2.2 Pseudorange Calculation

- The pseudorange is the apparent distance to a satellite: $\rho_i = d_i + c \cdot b_r$, where:

  - ρ_i: Pseudorange to the i-th satellite.

  - b_r: Receiver clock bias.

• •

3. ERROR SOURCES AND CORRECTIONS

3.1 Atmospheric Delays

1. **Ionospheric Delay:**

   – High-frequency signals slow down in the ionosphere.

   – Dual-frequency signals correct for this delay using:

   $$\Delta t_{iono} \propto \frac{1}{f_1^2} - \frac{1}{f_2^2},$$ where f_1 and f_2 are the frequencies of the signals.

2. **Tropospheric Delay:**

   – Modeled using temperature, pressure, and humidity data.

3.2 Multipath Interference

• Reflected signals create errors in timing and distance.

• Mitigated by advanced signal processing algorithms.

3.3 Satellite Orbital Errors

• Satellite positions are determined using Keplerian motion:

$$r = \frac{a(1 - e^2)}{1 + e \cos v},$$ where:

   – a: Semi-major axis.

   – e: Eccentricity.

– ^{v}: True anomaly.

• •

4. RELATIVITY AND GPS

4.1 Special Relativity

- The satellite's high velocity causes time dilation:

$$\Delta t_{special} = -\frac{v^2}{2c^2} \cdot t,$$ where v is the satellite's velocity.

4.2 General Relativity

- Weaker gravity at satellite altitude speeds up clocks:

$$\Delta t_{general} = \frac{GM}{c^2 r},$$ where:

- G: Gravitational constant.

- M: Earth's mass.

- r: Distance from Earth's center.

4.3 Combined Effect

- Net relativistic correction for satellites is approximately:

$$\Delta t_{relativity} = -38\,\mu s/day.$$

- Applied pre-launch to synchronize satellite clocks with Earth-based time.

• •

5. ERROR CORRECTION MODELS

5.1 Least Squares Adjustment

- Corrects over-determined systems of equations in trilateration.

- Objective: $\min \sum_{i=1}^{n} (\rho_i - d_i)^2$.

5.2 Kalman Filtering

- A recursive algorithm that integrates GPS data with external sensors (e.g., INS).

- Prediction step: $x_{k|k-1} = F \cdot x_{k-1|k-1} + B \cdot u_k$.

- Update step: $x_{k|k} = x_{k|k-1} + K_k \cdot (z_k - H \cdot x_{k|k-1})$, where:

   - K_k: Kalman gain.

   - z_k: Measured state.

. . . .

6. ADVANCED MATHEMATICAL TOOLS IN GPS

6.1 Fourier Analysis

- Used for signal processing to isolate GPS signals from noise.

- Fourier Transform: $X(f) = \int_{-\infty}^{\infty} x(t)e^{-j2\pi ft}\, dt.$

6.2 Matrix Algebra

- Represents systems of equations in compact form: $A \cdot x = b$, where A is the coefficient matrix, x is the solution vector, and b is the result vector.

6.3 Error Covariance Analysis

- Quantifies uncertainty in position estimates: $P = (A^T \cdot W \cdot A)^{-1}$, where P is the covariance matrix, and W is the weight matrix.

· ·

CONCLUSION

The mathematical foundations of GPS reveal the intricate interplay of geometry, physics, and advanced algorithms that underpin this transformative technology. By leveraging these principles, GPS achieves the precision and reliability required for its myriad applications, while ongoing advancements in modeling and computation continue to refine its capabilities. This mathematical framework ensures GPS remains a cornerstone of modern science and technology.

REFERENCE LIST OF TOOLS AND RESOURCES FOR GPS RESEARCH AND SECURITY

This reference list provides a comprehensive guide to the tools, software, and resources available for conducting GPS research, enhancing system security, and understanding GNSS technologies. It includes hardware, software, datasets, regulatory bodies, and publications widely used in the field.

• •

1. HARDWARE TOOLS

1.1 GPS Receivers

- **Trimble R12i GNSS Receiver:**

    – High-precision receiver for surveying and scientific applications.

- **u-blox ZED-F9P:**

    – Affordable, multi-band, RTK-capable GNSS module for research and development.

- **Garmin GPSMAP Series:**

    – Consumer-grade receivers for general-purpose navigation and testing.

1.2 Antennas

- **Tallysman GNSS Antennas:**

    – High-precision antennas supporting multiple constellations and frequencies.

- **Hemisphere GPS A222 GNSS Smart Antenna:**

    – Ideal for agricultural and machine control applications.

- **Choke-Ring Antennas (e.g., NovAtel):**

    – Used in scientific and surveying applications to minimize multipath effects.

1.3 Signal Generators and Analyzers

- **Rohde & Schwarz SMW200A Vector Signal Generator:**

    – High-end signal generator for simulating complex GNSS scenarios.

- **Keysight Technologies N9030A PXA Signal Analyzer:**

  – Advanced signal analyzer for identifying interference and analyzing GNSS signals.

- **Spirent GSS7000 GNSS Simulator:**

  – Versatile platform for testing GNSS receivers under controlled conditions.

1.4 Testbeds

- **GNSS Anechoic Chambers:**

  – Shielded environments for conducting interference-free GPS experiments.

  – Example: ETS-Lindgren chambers.

• •

2. SOFTWARE TOOLS

2.1 Signal Processing and Analysis

- **GNURadio:**

  – Open-source toolkit for building GNSS signal processing applications.

  – URL: https://www.gnuradio.org

- **MATLAB GNSS Toolbox:**

  – Provides algorithms and tools for GNSS signal generation, analysis, and receiver development.

  – URL: https://www.mathworks.com

- **LabVIEW GNSS Toolkit:**

  – Customizable platform for developing GNSS applications.

  – URL: https://www.ni.com

2.2 Simulation

- **Skydel SDX:**

  – Software-defined GNSS simulation platform.

  – URL: https://skydelsolutions.com

- **GPSTk (GPS Toolkit):**

  – Open-source library for GNSS data analysis and signal processing.

  – URL: https://www.gpstk.org

- **RTKLIB:**

- Open-source RTK positioning software for high-accuracy GNSS applications.

- URL: http://rtklib.com

2.3 Cybersecurity Tools

- **Kali Linux:**

  - Security-focused operating system with tools for testing GPS vulnerabilities.

  - URL: https://www.kali.org

- **Wireshark:**

  - Network protocol analyzer useful for examining GNSS-related data streams.

  - URL: https://www.wireshark.org

- **HackRF One:**

  - Software-defined radio platform for studying GNSS interference.

  - URL: https://greatscottgadgets.com/hackrf/

• •

3. DATASETS AND RESOURCES

3.1 GNSS Data Archives

- **International GNSS Service (IGS):**

  - Provides high-precision GNSS data and products for research.

  - URL: http://www.igs.org

- **National Geodetic Survey (NGS):**

  - Offers GPS data and tools for geodetic and mapping applications.

  - URL: https://geodesy.noaa.gov

3.2 Ephemeris and Clock Data

- **CORS Network:**

  - Real-time and archived GNSS data for differential corrections.

  - URL: https://www.ngs.noaa.gov/CORS/

- **SP3 Ephemeris Files:**

  - Standardized satellite orbit and clock files for GNSS processing.

3.3 Open GNSS Datasets

- **UNAVCO:**

  - Provides geophysical data, including GPS observations.

  - URL: https://www.unavco.org

4. REGULATORY BODIES AND STANDARDS

4.1 International Organizations

- **International Telecommunication Union (ITU):**

  - Allocates GNSS frequency bands and develops global standards.

  - URL: https://www.itu.int

- **United Nations Office for Outer Space Affairs (UNOOSA):**

  - Promotes international cooperation in GNSS development and usage.

  - URL: https://www.unoosa.org

4.2 National Authorities

- **Federal Communications Commission (FCC):**

  - U.S. regulatory body overseeing GNSS interference and licensing.

  - URL: https://www.fcc.gov

- **Ofcom:**

  - U.K. regulator managing GNSS-related spectrum usage.

  - URL: https://www.ofcom.org.uk

4.3 Standards and Guidelines

- **RTCA DO-229:**

- Standards for GPS Wide Area Augmentation System (WAAS) receivers.

- **ISO 19047:**

  - Standards for GNSS-based applications in intelligent transportation systems.

• •

5. PUBLICATIONS AND JOURNALS

5.1 Journals

- **GPS Solutions:**

    – Publishes research on GNSS theory, applications, and innovations.

    – URL: https://link.springer.com/journal/10291

- **Journal of Navigation:**

    – Covers advances in navigation technologies, including GNSS.

    – URL: https://www.cambridge.org/core/journals/journal-of-navigation

5.2 Books

- **"Understanding GPS/GNSS: Principles and Applications" by Elliott D. Kaplan and Christopher J. Hegarty:**

    – Comprehensive guide to GNSS theory and practice.

- **"GPS for Land Surveyors" by Jan Van Sickle:**

    – Focuses on the practical aspects of GPS usage in surveying and mapping.

• •

CONCLUSION

The tools and resources listed here provide a robust foundation for researchers, developers, and security professionals working with GPS and GNSS technologies. By leveraging these resources, you can enhance your understanding of GPS, address its vulnerabilities, and innovate solutions for its secure and reliable use in critical applications.

FURTHER READING ON GNSS POLICIES AND ADVANCEMENTS

The rapid evolution of Global Navigation Satellite Systems (GNSS) has spurred significant advancements in technology, policy, and applications. This further reading section provides recommended resources for understanding the global policies, technological progress, and future directions in GNSS. These resources span books, research papers, online repositories, and institutional publications, providing a comprehensive base for deeper exploration.

• •

1. BOOKS ON GNSS POLICIES AND TECHNOLOGICAL ADVANCEMENTS

1.1 Policy and Regulation

- **"Satellite Navigation Systems: Policy, Commercial and Technical Interaction" by Michael J. Rycroft**

  - Explores the intersection of policy, commercial applications, and technical innovations in GNSS.

  - Focuses on international cooperation and regulatory challenges.

- **"The Politics of Space Security: Strategic Restraint and the Pursuit of National Interests" by James Clay Moltz**

  - Examines the geopolitical implications of space-based systems, including GNSS.

  - Discusses the militarization of GNSS and its impact on global security.

• •

1.2 Technological Advancements

- **"Understanding GPS/GNSS: Principles and Applications" by Elliott D. Kaplan and Christopher J. Hegarty**

  - A comprehensive technical resource covering the

design, development, and applications of GNSS.

- Includes discussions on advanced GNSS receivers, multi-constellation systems, and future trends.

- **"GNSS Software Receivers: Implementation, Simulation, and Testing" by Kai Borre et al.**

    - Focuses on the software-based implementation of GNSS receivers, enabling flexible research and testing.

    - Explores simulation and real-world testing scenarios.

• •

2. RESEARCH PAPERS AND ARTICLES

2.1 Policy and Governance

- **"The Outer Space Treaty at 50: An Enduring Basis for Cooperation" by Paul Meyer**

  - Published in *Disarmament Times*, this paper reviews the relevance of the Outer Space Treaty in regulating GNSS operations and ensuring peaceful use.

  - URL: https://www.unoosa.org

- **"Spectrum Management for GNSS: Challenges and Opportunities" by ITU-R Study Group 4**

  - Explores the critical role of spectrum allocation in GNSS performance and interference prevention.

  - Published by the International Telecommunication Union (ITU).

• •

2.2 Technological Trends

- **"Multi-GNSS Constellation Benefits for Precision Applications" by Guo Chen et al.**

  - Published in *GPS Solutions*, this paper highlights the advantages of combining multiple GNSS constellations for high-accuracy positioning.

  - URL: https://link.springer.com/journal/10291

- **"Emerging GNSS Threats: A Comprehensive Analysis of**

Jamming and Spoofing" by Christopher J. Hegarty

 –   Published in *IEEE Transactions on Aerospace and Electronic Systems*, this article explores vulnerabilities and mitigation strategies in GNSS.

• •

3. INSTITUTIONAL PUBLICATIONS

3.1 International Telecommunication Union (ITU)

- **Publications on GNSS Spectrum Management:**

    - Explores regulatory frameworks for GNSS frequency bands and international coordination efforts.

    - URL: https://www.itu.int

3.2 United Nations Office for Outer Space Affairs (UNOOSA)

- **GNSS in Developing Countries:**

    - Highlights GNSS applications for sustainable development and policy recommendations for emerging economies.

    - URL: https://www.unoosa.org

3.3 European GNSS Agency (GSA/EUSPA)

- **Market Report on GNSS Trends:**

    - Provides insights into GNSS market dynamics, technological advancements, and policy developments.

    - URL: https://www.euspa.europa.eu

· ·

4. ONLINE RESOURCES AND REPOSITORIES

4.1 GNSS Monitoring and Analysis

- **International GNSS Service (IGS):**

  – Offers GNSS data, products, and tools for research and operational use.

  – URL: https://www.igs.org

- **Navipedia:**

  – An open-source knowledge base maintained by the European Space Agency (ESA), covering all aspects of GNSS.

  – URL: https://gssc.esa.int/navipedia

• •

4.2 Industry Reports and White Papers

- **Trimble GNSS Technology Reports:**

  – Insights into the latest advancements in GNSS receivers and applications.

  – URL: https://www.trimble.com

- **Rohde & Schwarz GNSS Security White Papers:**

  – Detailed discussions on GNSS vulnerabilities and countermeasures.

–   URL: https://www.rohde-schwarz.com

. . . .

5. CONFERENCE PROCEEDINGS

5.1 Institute of Navigation (ION) GNSS+

- **Annual Conference:**

    - A premier forum for GNSS researchers, featuring the latest papers on GNSS policies, technologies, and applications.

    - URL: https://www.ion.org

5.2 IEEE/ION Position, Location, and Navigation Symposium (PLANS)

- **Key Topics:**

    - Advances in navigation technologies, including GNSS modernization and multi-sensor integration.

    - URL: https://ieeexplore.ieee.org

• •

6. EMERGING TOPICS IN GNSS

6.1 Multi-Constellation Integration

- **"Interoperability of GNSS Systems: Challenges and Solutions" by EU GNSS Agency**

  – Discusses efforts to harmonize signals across GPS, Galileo, BeiDou, and GLONASS for improved user experience.

6.2 Quantum Technologies

- **"Quantum Clocks and Their Role in Next-Generation GNSS" by NIST**

  – Explores how advancements in quantum clocks can enhance GNSS timing and positioning accuracy.

6.3 Space-Based Augmentation Systems (SBAS)

- **"Enhancing GNSS Performance with SBAS: Current Trends" by RTCA**

  – Reviews the impact of SBAS systems like WAAS and EGNOS on aviation and precision agriculture.

• •

CONCLUSION

The recommended resources provide a thorough foundation for understanding the policies and technological advancements shaping the future of GNSS. These materials are valuable for researchers, policymakers, and professionals looking to deepen their knowledge and contribute to the evolution of GNSS technology in an increasingly connected world.

www.ingramcontent.com/pod-product-compliance
Lightning Source LLC
Chambersburg PA
CBHW071450220526
45472CB00003B/745